Praise for *The Farmer's Office*

I keep telling people, starting your own small farm is not easy; to make it work, you need—perhaps more than anything else—to learn strong business skills. This is the key element that *The Farmer's Office* brings to the table; in it Julia Shanks shares solid advice about how to make your farm work financially. A definite must-read for any serious small-scale grower.

— Jean-Martin Fortier, author, *The Market Gardener*

Julia Shanks has distilled essential lessons that go beyond business planning to the financial know-how of operating a profitable farm. I will recommend her book to our clients.

—Dorothy Suput, Executive Director, The Carrot Project

Managing books and accounts is one of the weakest links I see with farmers today, and *The Farmer's Office* comes at a perfect time to help. There's so much talk about production and even marketing, but not enough on how to stay organized and manage the books. *The Farmer's Office* will be a great asset to farmers of all shapes and sizes.

—Curtis Stone, author, *The Urban Farmer*

Julia Shanks takes all the key topics to cover and lays them out in a logical order. She then uses stories and examples to bring home why it is so important. This book will become the go-to business guide for many, many, many farmers!

—Jonathan W. Jaffe, VP/Farm Business Consultant, Farm Credit East

The Farmer's Office hits the mark in addressing the business needs of farmers everywhere. Julia Shanks shares her business experience and real world financial training in a meaningful and farmer friendly format. While a farmer's footprints may be the best manure, *The Farmer's Office* is a key asset to the farm's bottom line.

—Richard Wiswall, farmer and author,
The Organic Farmer's Business Handbook

This is the real story of starting and managing a farm, albeit the unsexy one that can lead to long term thrive-ability. Julia Shanks demystifies managing the business of farming one chapter at a time. This book is just what so many of today's new farmers need to face the challenges of agriculture and thrive!

— Sara Dent, cofounder & coordinator, Young Agrarians

I don't know anyone who got into farming because they liked business; but anyone who wants to keep farming needs basic business skills, including record keeping. [This book] will help everyone from beginners to experienced growers who need to understand the business of farming. A must-have for anyone running, or thinking of running a farm business.

—Andrew Mefferd, Editor, *Growing for Market* magazine

Julia Shanks has helped our farm immeasurably. We are better at growing diverse foods than at figuring out if specific crops make us money, break even, or lose money. Indeed, I am sometimes even afraid of figuring out detailed profit realities all by myself. Julia's economic analyses and insights into how farms work have been really empowering, and *The Farmer's Office* is simple, easily digested, and brings to life even the sometimes scary subject of how to run a successful farm business. This book is a must-have for all new farmers, and even we grizzled experienced growers will gain hugely from a cover-to-cover read!

—Brett Grohsgal, Even' Star Organic Farm

Julia has a uniquely comprehensive skill set that is especially beneficial to the Farmer Entrepreneur. Her deep understanding of farmers, business, and the business of farming make *The Farmer's Office* both highly functional and extremely informative.

—Susan Parke-Sutherland, Wingate Farm

THE
FARMER'S
OFFICE

THE
FARMER'S
OFFICE

TOOLS, TIPS AND TEMPLATES
TO SUCCESSFULLY MANAGE A GROWING
FARM BUSINESS

JULIA SHANKS

new society
PUBLISHERS

Cover design by Diane McIntosh.
All cover images © iStock. Chapter image © Raven/Adobe Stock.

Printed in Canada. First printing September 2016.

Inquiries regarding requests to reprint all or part of *The Farmer's Office*
should be addressed to New Society Publishers at the address below.
To order directly from the publishers, please call toll-free (North America)
1-800-567-6772, or order online at www.newsociety.com

Any other inquiries can be directed by mail to:

New Society Publishers
P.O. Box 189, Gabriola Island, BC V0R 1X0, Canada
(250) 247-9737

LIBRARY AND ARCHIVES CANADA CATALOGUING IN PUBLICATION

Shanks, Julia, author
The farmer's office : tools, tips and templates to successfully manage
a growing farm business / Julia Shanks.

Includes bibliographical references and index.
Issued in print and electronic formats.
ISBN 978-0-86571-816-6 (paperback) — ISBN 978-1-55092-610-1 (ebook)

1. Farm management. 2. Farms, Small—Management. I. Title.

S561.S53 2016   630.68   C2016-903937-4
                                       C2016-903938-2

New Society Publishers' mission is to publish books that contribute in
fundamental ways to building an ecologically sustainable and just society, and to do so with
the least possible impact upon the environment, in a manner that models that vision.

To all the farmers
who have demonstrated that it's possible
to be both financially and ecologically sustainable.
You give us faith that we can build a more
sustainable and just food system.

Contents

Gratitude

This section of a book is typically titled "Acknowledgements." But that word does not do justice to my appreciation to all the friends, colleagues and mentors who have inspired and taught me over the years and enabled me to write this book.

Like most farmers, Brett Grohsgal constantly moves—feeding chickens, weeding, planting and harvesting. When I visit him, our time to connect and talk is in the fields; I work alongside him, and we talk about his business goals and struggles. If my clients struggle, he shares his insights. Through years of conversations and friendship, I gained firsthand insight into the joys and challenges of farming. I began to see the value in managing by the numbers. Who knew that a friendship with a farmer could be so educational? And for the friendship and education, I am most grateful.

I first started creating Excel templates for farmers while consulting with The Carrot Project. Their forward thinking regarding the needs of young farmers started my process in developing more tools. I'm grateful to Dorothy Suput, the Executive Director, for the initial guidance (and continued friendship). As I expanded the resources, Genevieve Goldleaf and Noelle Fogg provided invaluable assistance that enhanced and supported my work.

My understanding of accounting and business management came through a multitude of connections and experiences. Ginny Soybel taught me accounting at Babson College and mentored me throughout my time there both as a student and eventual instructor of accounting. Roman Weil and Fred Nanni also coached and mentored me.

Managerial accounting is best learned through real-world experiences, and I'm grateful to have worked with Denise Chew, Michael Staub, Craig Richov and Melissa Adams.

Many friends and colleagues read chapters, provided suggestions, asked questions and generally helped me improve the book. I'm grateful for the help of Sarah Andrysiak, Dan Banks, Meghan Bodo, Denise Chew, Katherine Collins, Annie Copps, Myrna Greenfield, Alex Loud, John Paskowski, Lisa Sebesta, Hershel Shanks, and Judith Shanks.

It takes a village to write and publish a book, and New Society could not have provided a better village of editors and marketers. I so enjoyed working with the entire team, most especially Ingrid, Judith, Sue, EJ and Sara.

Finally, to all the farmers and entrepreneurs I've had the pleasure to work with over the years. I owe a debt of gratitude for all you do to support a sustainable food system. I've learned so much from you, and hopefully you've learned something from me too.

Foreword

by Richard Wiswall

Odds are you picked up this book because you've realized that success in farming is more than all the tasks that go into raising plants and animals and marketing them. Farming also has a side that is business; a side that cannot be forever ignored. Yes, farming is indeed a business. Sorry to burst your bubble if you thought otherwise. Welcome to one of the most unglamorous and avoided topics in farming. Oh, but so important.

Ask any farmer why they are attracted to farming, and I bet it wasn't because of the desire to learn about pro forma balance sheets, cash flow projections, or income statements. I don't think that farmers have a genetic defect in regard to business learning, but the words "farming" and "business" seem to separate like oil and vinegar.

I've farmed full time for the last 35 years, and I still remember the moment when I got pulled into the business world. When I started farming I thought, like many other farmers, that if I just worked hard enough, everything would turn out all right. Sometimes that happens, but often aided by dumb luck. Working harder and harder has its limits, and doesn't always work. And overwork can lead to frustration, burnout—and possibly an early exit from farming altogether. Farming and the 'money thing' need to be reconciled.

The good news is that, once I accepted the idea that farming is indeed a business, I became a better farmer. I began farming smarter, not harder. Time spent on the business became more of a priority. The business side of farming is now second nature; in fact, I no longer see any separation between the concepts of "farming" and "business". They are one, plain and simple.

Let's back up a little. Why do you keep records at all? Because conventional wisdom says to? Because I say to? No. Because you have to file taxes?

Yes, partially so. But ultimately, the reason to track numbers is to better manage your business; to shine a light on the inner workings of your farm so you can make better decisions to work more efficiently and make more money. It's farming's win-win.

The Farmer's Office is the light that illuminates these concepts of business for farmers. Julia has firsthand knowledge of the subject from her years in her own business, and writes in an easy to read style. Real stories from real farmers, from her many years of helping farm businesses, illustrate various issues and applications when managing a farm. In exploring a topic, *The Farmer's Office* poses all the right questions and provides clear answers to them. Julia covers all the bases in a step-by-step manner that will help beginning and seasoned farmers alike. This book is for you, for every farmer.

What are you waiting for? Go ahead and take the plunge. What's the downside? You lose a few hours of your time? Short circuit your cerebral cortex with smoke coming out of your ears? The upside is that your farm will thrive with new knowledge of business principles, now, and for long into the future.

Happy Farming.

— Richard Wiswall,
Owner, Cate Farm and author of
The Organic Farmer's Business Handbook

Preface

People often ask me how I got into this line of work—working with farmers to help them with business planning and teaching basic accounting. My career certainly didn't start in accounting nor in agriculture. It started with food.

Soon after college (and after a small career detour), I went to culinary school. My first job out of cooking school was at one of the original "farm-to-table" restaurants. It was the early '90s before farm-to-table was even a concept. Brett Grohsgal was the chef, and I was a line cook. We maintained our friendship when he moved to southern Maryland with his wife to farm on newly purchased land and I moved to Boston. Our friendship grew on his farm. I visited once or twice a year, and I always worked alongside him in the fields—it was the only way to have a conversation with him. I helped weed, slaughter chickens, harvest watermelons and pick okra. We mostly commiserated about entrepreneurship—I was running a catering business, and he was running his farm. In all our conversations, I learned about farming, too.

After 10 years of cooking, I was ready for a change and went to Babson College to earn an MBA in entrepreneurship.

In my first year of business school, we were presented with the following problem:

A farmer in Iowa owns 45 acres of land. She is going to plant each acre with corn or potato. Each acre planted with potatoes will yield $200 profit; each with corn yields $300 profit. Each acre of potato requires 3 workers and 2 tons of fertilizer. Each acre of corn requires 2 workers and 4 tons of fertilizer. One hundred workers are available, and 120 tons of fertilizer are available. What is the optimal mix of potatoes and corn that the farmer should plant to maximize profits?

I plugged all the numbers into an Excel spreadsheet, opened the solver box and clicked "solve." The computer spat out the answer: 20 acres of each.

The computer modeling fascinated me. The following week, spring break, I headed down to Brett's farm. I wanted to see if I could help Brett maximize his profits. We calculated all the costs to get 10 of his crops into the ground, out of the ground and to his customers. We flipped through his sales book to estimate his annual yields (based on previous years) and the profit. I created an elaborate Excel spreadsheet, organized the numbers and clicked on "solve." The goal was to determine the optimal mix of crops Brett should sell to maximize his profits. The answer wasn't particularly helpful: it said that Brett should only grow okra and sweet potatoes. I didn't need a business degree to know that a farm that only sells okra and sweet potatoes wouldn't have many customers. But through the course of this exercise, Brett realized he was making 12 cents for each case of tomatoes and needed to raise his prices.

This was the first time I truly understood the power of the numbers. Too often, we price our products based on what the competition is doing. We tell ourselves, "This is what the market will bear." Without knowing his true costs, Brett would continue to hobble along in business. Knowing his true costs enabled him to evaluate his options and make an informed decision:

- Raise his prices
- Discontinue tomatoes or
- Continue selling tomatoes at a discounted price as a way to attract customers.

If he raised his prices, then he would need to communicate to his customers the superiority of his tomatoes and why they commanded a higher price. He could stop selling tomatoes, but worried that he wouldn't have the appeal at the farmers market without them.

What started as a random exercise to help a friend led me down the path of working with farmers. After graduation, I went back to Babson to teach accounting and started my consulting practice. My mission was to help farmers (and other food-system entrepreneurs) grow their businesses and manage their finances.

As I started working with entrepreneurs, I discovered where they really needed the support. I would get comments like, "I don't understand why I don't have any money in my bank account." Or, "I want to grow my business but I'm not sure where I should focus my energies." Every business is different, and there's no pat answer. With each client, I would ask to see their books. Some used QuickBooks, some just used Excel spreadsheets. But the systems were set up for balancing their checkbooks and filing taxes, not for making informed decisions based on facts. It became apparent that I needed to first coach clients on accounting and bookkeeping systems before we could tackle the bigger questions of cashflow management and growth strategy.

The lessons of good bookkeeping came into sharp focus as I worked with clients teetering on the edge of solvency. They had fallen behind on taxes, with their vendors, borrowed to the hilt from banks and family, and could not figure out how to get themselves out of the hole. These clients didn't have the luxury of time—we couldn't spend three months straightening out their books, getting clean numbers into their systems so we could do effective analysis. Decisions had to be made quickly to keep them from going under.

There is only so far an entrepreneur can manage her business without having a solid grasp of the financials. Some farmers get lucky—they have a good intuition of what things cost, enter a solid market at the right time and are wildly successful. Most of us aren't that lucky. We need to make thoughtful decisions and spend our money wisely.

It's clear: the only way to make informed decisions about your business is to have a solid grasp of the numbers. And to have a solid grasp of the numbers, you need a basic understanding of accounting and a solid bookkeeping system.

The Farmer's Office evolved from the desire to help as many farmers as possible to be financially viable, and to share the lessons learned from other farmers. On a personal level, more successful small farms means more delicious food to eat, a cleaner environment and a more robust economy. It benefits us all to ensure our local farmers succeed.

Wishing you continued success as you grow your business!

—Julia Shanks

Disclaimer

The accounting profession uses some words and terms with precise meanings and often refers to generally accepted accounting principles (GAAP). Accounting courses use the technical meanings and refer to GAAP, as do public companies when they report to the Securities and Exchange Commission. For small business owners, the meanings of these terms and practices can be confusing and burdensome.

Financial and managerial accounting typically require several years for the student to master. I want to make accounting approachable and easy to digest within a few months. In some places, I take a simplified approach and may fudge some of GAAP. Some of the "simplifications" are fine for business planning and analysis, but need more precision for taxes. I will let you know when you should consult an accountant.

I am neither a tax accountant nor a tax expert. When it comes to filing income taxes, as well as remitting sales and payroll taxes, I strongly suggest you consult with a tax accountant and payroll service provider.

All of the stories and examples derive from real farmers and entrepreneurs with whom I have worked. The numbers are mostly a true reflection of their real business experiences. I have changed some stories and details to protect their privacy.

How to Use This Book

There is definitely a process to proper bookkeeping and business management, but every business is different and needs different tools at different times. While I wrote this book with the business cycle in mind, it may not be the cycle of *your* business.

Whatever your stage of business, I recommend that you start with Chapters 1 and 2 ("Introduction—Why Bother" and "Building the Foundation"). This sets the foundation for good accounting practices. Depending on your business's stage of development, you may prefer to skip some chapters. Here's a decision tree to help you decide which chapters to read next.

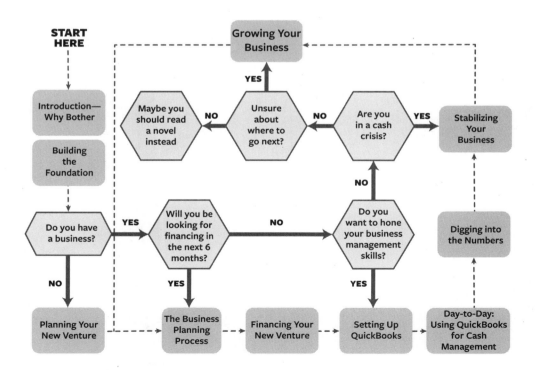

There are a few chapters/sections that get pretty deep into the numbers. Save these chapters for when you can actually follow along with your own business:

- The tutorial on setting up QuickBooks is best read when you're actually setting up your QuickBooks file.
- The tutorial on Creating Financial Projections will make more sense if you're working through your own.

Finally, if you learn better by listening than reading, you can watch videos and webinars to complement your learning: juliashanks.com/video-tutorials/

CHAPTER 1

Introduction—
Why Bother

How's business?

If you're like most farmers, you'd answer with a recount of the recent rains, the arugula harvest or the flea beetle you're battling. It wouldn't be an overview of your profitability.

After all, you decided to become a farmer because you love being outside, working the land and making a difference in the way we eat and farm. Of course, you want to earn a decent living to support yourself and your family.

When you decided to become a farmer, you also became an entrepreneur and a business person. In order to be ecologically *and* financially sustainable, you must understand the basics of accounting and bookkeeping. Good bookkeeping can give you the information you need to improve your profitability, make good investment decisions and manage your debt. And you will have what you need to write a business plan, which you may have to do every once in a while. In other words, good business management helps you make money.

Really?! Yes, really.

Let's face it, accounting is not sexy, and most people dread it more than the dentist. Every so often, I have a client who discovers the power of bookkeeping and the story that numbers can tell about a business. This happened for Laura Meister of Farm Girl Farm. I was curious, what sparked her interest and how could her story inspire other farmers to see the value in numbers? I asked her, "Why Bother?" Her response was pretty powerful!

How Much for How Much?   *by Laura Meister*

When I started growing vegetables ten years ago, it was all I could do to keep up with the start-up math: how many square feet in an acre again? How many CSA[1] members do we think we can sign up in our first season? How much food do they expect in a box? How many weeks are we serving them? So, then, how much do we grow? How much can we grow? And how many seeds does that mean? And when I finally had all those numbers banged out, I was nearly done in by the Fedco catalog—now I've got to convert ounces to grams? Are you kidding me?

Once all that was more or less behind me and the arrival of spring forced my attention to the real playing field—the actual field—I abandoned my desk entirely. I thought I'd made my plans well enough, and if I now rode the rollercoaster with my white knuckles gripping the bar until Thanksgiving, I'd surely have some money in my pocket to show for all this sweat.

I worked hard. Really hard. Really goddamned hard. You know how hard I worked because you work that way too. I barely slept. I lost my business partner because it turned out this kind of hard work was not what she'd had in mind. Although I made every rookie mistake in the book, I managed to wrestle some produce from the ground and feed my 40 CSA members. I even had some surplus so I started calling scary chefs who turned out to be less scary than I thought and wanted to buy what I was selling. So when the snow finally flew that fall, I thought I'd had a pretty successful season.

But I had no money in the bank and a giant credit card bill from start-up expenses that I'd never repaid. So I got a job stocking produce at our local Co-op and worked until late spring the following season. Thank god for that job, but I thought I'd be able to use the winter months to plan the next season and go to a few yoga classes to heal my back. I started Season Two behind the eight ball, as my mother says.

One thing was clear to me—my farming days were numbered. I could not afford to work another season of blood, sweat and tears, only to find myself perhaps deeper in debt than after Season One. I knew I had to do something, and I knew I couldn't possibly work harder, so I was going to have to work differently. In my neighborhood, an organization to support farmers called Berkshire Grown sponsored a business class, Tilling the Soil of Opportunity, a 10-week course designed for new farm operators. I signed up.

We were invited to bring our financial records from previous seasons. I had none. Not a receipt, not an invoice. The last thing I'd written down was where I was going to plant the tomatoes. So I started from scratch. "It's ok," my instructor said. "Guess." With his coaching, I guessed at every single expense for the coming season, by month, and potential revenue from all my potential customers and income sources. Tedious does not begin to describe the process. But at the end, I had a cash flow plan. I used that plan like a roadmap for Season Two. Whenever a decision came at me, instead of wavering, hemming and hawing, I simply looked at my plan—a friend wanted to come work for me for the season, could I afford her? (No.) Should I buy the tractor that sounds exactly like the one I need? (Not now.) How many tomato stakes can I buy this week, and how many if I wait until next week? (100 now and 100 next week). The cash flow plan was my North Star.

At first I thought it was an amazing coincidence that my real numbers turned out to be so close to what I'd guessed. Later I realized the enormous impact of having a plan—I didn't go wildly over my projected expenses because I consulted the plan before I spent money. If my income wasn't adding up to the projections, I made a couple of more phone calls, and sold a few more turnips until I was where I was supposed to be.

There have been good seasons and bad seasons since those first two—some years I had a solid plan, some years I had tomato blight or a hurricane. After those disasters, I was very resistant to face my numbers and make a good plan for the following season; it was just too depressing. And after a few years of avoiding the numbers, I was back in a hole of debt. This past winter, I faced my fears, rolled up my sleeves and got way down and dirty with the details.

I know now that ignoring the numbers and expecting to be able to continue to do the work I love is tantamount to not eating and expecting to be able to work through the day. If I don't take care of the numbers, I will be back at that produce-stocking job in a heartbeat.

> Believe it or not, I now look forward to my time with the numbers.
> I used to think they were black and white and boring, but they are
> magical and powerful teachers, and they are the keys to the kingdom.
> Here's to our calculators and a successful growing season!

At a very base level, you need to manage your books so that you can file your taxes at the end of the year. If you need a loan to purchase a new tractor, the bank will want a profit and loss (P&L) statement, a balance sheet and a statement of cash flows. And if you're applying for a grant, you'll need a business plan.

Beyond that, good bookkeeping can actually help you reduce costs (even if you hire a bookkeeper) and keep you out of trouble. There's so much richness in the numbers that you do yourself a disservice to ignore them.

Understand What Makes You Money (and Why)

Do you know which farmers market is most profitable for you? Do you know if you're making money on your chickens? You probably have a gut instinct. But how bummed would you be if you realized that you were actually losing money on the meat birds? While you may be tracking that more money is coming into your business than is leaving (i.e., you are running a profitable business), it could be that some of your products are losing money. If you don't track your revenues and expenses effectively, then you don't know for sure.

Brett Grohsgal of Even' Star Organic Farm in Southern Maryland launched his business in 1997, thinking he would be the "Tomato Man," and grew 60 varieties that were truly vine ripened. After four years in business, things were going well; he started diversifying his crops and selling through a CSA model, but he knew he could do better. In February 2005, as he plotted his crop plan, we sat down with his books. We went through every expense and every sale as it related to each crop. We figured how much time he spent propagating seeds, sowing, weeding, harvesting and sorting. We calculated the cost to produce a case of tomatoes that he sold for $30. When he factored in all the labor and production costs, he was netting only 12 cents a case![2] Cucumbers, okra and sweet potatoes, on the other hand, were cash cows. He knew that no amount of tomato sales would be sufficient to run a profitable farm; if he wanted to stay in business, he needed to further diversify his vegetable offerings and limit tomato production.[3]

By understanding what makes you money, you can focus your current operations to maximize profitability and develop strategies for growth.

Invest in Your Business

Can you afford a greenhouse, or to hire employees? Is that tractor in your budget?

Hannah and Ben Wolbach of Skinny Dip Farm were overworked and understaffed. They barely hired any help because they "had no confidence in their ability to forecast for even one season ahead." They weren't sure if they could afford the added expense.

By creating a cash flow budget, they had the confidence to grow their business and hire staff because they could project the impact of payroll expense on their profitability.

Plan for Growth

What's the best way to grow your business? Should you expand current operations, or diversify your offerings? Can you afford the loan that will help finance your plans?

Wingate Farm in New Hampshire wanted to devise a strategy to grow its business. It was raising meat birds, laying birds and vegetables. After their first year, owners Susan and Olivia had an $8,000 profit. They were pleased to have earned a modest profit, but also knew they needed to do better to grow a sustainable business. They questioned where to focus their attention. Similar to the analysis Brett did, Susan and Olivia combed through their detailed record keeping and discovered that, after accounting for labor, they were earning only 10 cents a bird! Meanwhile, the eggs were netting $1.15 a dozen. These facts informed their strategy—both in terms of focusing on egg production and reconfiguring the meat-bird operation to be more profitable.

Wingate Farms demonstrates how effective bookkeeping and understanding your numbers can be in helping you develop a growth strategy—by providing information about the profitability of specific products.

Good bookkeeping can also help you understand whether a new opportunity makes sense and if you can afford the loans that may be needed. Will that new tractor improve efficiencies enough to increase profitability even after you repay the loan that's required? Does it make sense to restore an old cider press?

With a growth strategy in mind, you need a plan to get there. What investments do you need to make, and how will you finance them?

Kate Stillman of Stillman's Quality Meats wanted to build her own meat processing facility. There's no question that having the capability to slaughter and butcher her own chickens, as well as butcher her own quadrupeds, would improve the quality of her product, increase efficiencies and help differentiate her product from an increasingly competitive local-meat landscape. The real question was, could she afford it? If she took out a loan to build the facility, could she generate enough profit to pay off the debt? Would the improved efficiencies of the processing facility actually reduce her costs?

By using sound financial forecasting, she decided the answers to these questions were yes, yes and yes. And instead of trying to save enough to finance the processing facility, she opted for bank loans and a USDA grant. The banks and USDA wanted to see the numbers and the story behind the numbers (that is, the business plan). With her books in good shape, she had what she needed to apply for the financing.

Plan for Slow Periods

Do you know how to plan for the inevitable ebbs and flows of business and cash? During slow periods (like August or January), cash can be tight. If you don't plan properly, and cash is tight, a few things can happen:

- You charge things to your credit card or take out short-term loans; both incur interest.
- You take unnecessary loans or even high-interest rate loans.
- You fall behind on paying your vendors, and they apply finance charges, demand payments in advance of delivery and/or refuse doing business with you altogether.

The unnecessary interest and finance charges can be avoided with proper planning.

A few summers ago, I visited a client, Roberta. She farms outside Washington, DC, and sells primarily to the DC market. During her peak production in August, her customers abandon the city for vacation destinations, and her wholesale restaurant clients decrease their order sizes. Therefore, in August, she has to be especially careful about cash flow. During our visit, I overheard her asking a worker to wait a few days to cash his paycheck. Given that she knows August will be slow, she can plan for it by understanding her cash shortfall, and budgeting for it.

Avoid Out-of-control Debt

I started working with Laura Meister from Farm Girl Farm not because she wanted to learn bookkeeping and budgeting, but because debt overwhelmed her. She hadn't budgeted properly for a couple of years, borrowed more than she could afford to pay back and got herself in trouble. In order to figure out the right strategy for Laura, we first had to determine the cause of the mounting debt. Was her business financially sustainable outside the burden of debt? If not, then the increase in debt was resulting from poorly managed operations or a business model that could not succeed. Good accounting records can help her figure out which.

If her base operations were solid, then we needed to dig deeper still. The increasing debt could be because she wasn't generating enough cash to pay it down. Or, that she paid debt too aggressively when she had cash, and then got herself in trouble again during the slow months.

We were able to get her books in order so that we could see:
- Her base operations were profitable.
- The amount she could afford to pay down and when.
- Where she could trim expenses to improve profitability (thereby improving her cash position).

Armed with this knowledge, she could create a plan to streamline operations and pay down debt without compromising her cash flow.

Keep Your Eye on the Prize

For many, the end goal in running a successful farm business is to feed the local food system and enjoy a sustainable lifestyle. To get there, you must understand what makes you money, have an understanding of the best way to grow and have a sound plan before taking on debt. It's about understanding the nature of your business with all its seasonal variances.

As Susan from Wingate Farm describes it:

So much of farming on a smaller scale is about reacting to situations that require immediate attention. It's easy to spend all day running around putting out fires. Amidst all that chaos, it's very easy to lose sight of the bigger picture. Taking the time to think through the big picture when you actually have the brain space/energy is the only way to farm with intention. Then you have a map for your farm and your business plan guides you through the chaos. It allows you to be proactive, instead of reactive.

The goal of this book is to give you the tools to be proactive in how you launch, grow and manage your business. And should you get yourself in trouble and into a cash crisis, you'll have the skills to evaluate your business and correct your course. This book will cover:

- The three primary financial statements (income statement, balance sheet and statement of cash flows)
- The underlying principles of the three financial statements
- How to use the numbers and information in the financial statements to better manage and grow your business
- How to set up your bookkeeping system so you can get the information you need
- Writing a business plan and creating financial projections for a loan or grant
- The underlying growth strategies of the business plan
- How to navigate a cash crisis.

I know business management, accounting and numbers can feel scary. But we'll chunk it down and go step by step. After weaving your own experiences with the concepts in this book, you'll be able to manage both your fields and your office.

Ready? Let's go!

Notes

1. Community Supported Agriculture (CSA) is a business model built upon the relationship between farmer and consumer. CSA farms receive customer payment in advance of the harvest in exchange for weekly distributions of produce during the growing season. This provides the farmer with much-needed

cash at the beginning of the season and offers the customers a connection with the farmer and the assurance of fresh local produce.

2. Nanni, Alfred J., Pachamanova, Dessislava and Shanks, Julia, Even' Star Organic Farm (July 27, 2007). AAA 2008 MAS Meeting Paper. Available at SSRN: ssrn.com/abstract=1003386

3. He still grew tomatoes as it was a way to lure customers to his farm. He also knew that selling only cucumbers, okra and sweet potatoes was not a viable option from a marketing perspective.

CHAPTER 2

Building the Foundation: The Financial Statements and Basic Accounting

Whether you're writing a business plan to launch a new venture, or navigating the day-to-day operations of your farm, the financial statements organize your numbers and tell a story of how your business is doing and where it's going. The three primary financial statements are:

- the income statement[1]
- the balance sheet
- the statement of cash flows.

For an overview of basic accounting, watch the webinar "Basic Accounting for Farmers" at: juliashanks.com/video-tutorials/

They can represent the past (historical) or the future (projections). Each financial statement represents a summary of various transactions that occur throughout your business and operations, and present different nuggets of insights about your business, your financial position, and your cash flow.

As the old saying goes, "You must crawl before you can walk." Before we can talk about creating financial statements or even setting you up in QuickBooks, we need to make sure we're speaking the same language: accounting-ese. These terms and concepts appear throughout business: whether you're talking with your banker or accountant or applying for a loan or grant. It's important to understand the common meanings.

If you're new to financial statements and basic accounting, you may find this language confusing and intimidating. Don't worry if you don't have all this committed to memory by the end of the chapter. As you see the concepts being put into action throughout the book, and as you begin using them yourself, you will feel more comfortable with them. Of course, you can refer back to this chapter and the videos as needed.

The Income Statement—A Summary of the Operations

For more details, see these videos: juliashanks.com/the-income-statement/ and juliashanks.com/multi-step-income-statement-in-depth/

When to Use

- For the business owner: to understand the profitability of your business, as well as the nuances of your different product lines and revenue streams (historical)
- For potential funders, entrepreneurs and partners: to project the profitability of a new venture (projections)
- For the government: to file your taxes (historical).

Key Features

- The income statement is a detail of the business activity over a period of time, usually a month, quarter or year.
- The income statement details the activities directly related to the operation of your business (selling produce, paying employees), as well as indirect activities (money earned from renting land or interest income).
- The income statement shows your total revenue, total expenses and net profit.
- The income statement does not track cash flow.

The income statement presents a summary of everything you earn through the course of your business. It includes selling products and/or services, as well as the cost associated with running your business: what you spent to purchase your seeds and soil amendments, to pay your employees, for advertising, rent, and so on.

More academically, an income statement provides information about businesses' operations in terms of profitability over a given period of time. It is broken down into two major sections that distinguish the farm's primary operations from the secondary transactions, which are divided by "the line."[2]

Here's a sample income statement from Stone Hill Farm. We'll talk more about them in chapter 4, The Business Planning Process.

The income statement is divided into general sections:

1. **Sales:** revenues of the farm for the past year or accounting period (which could also be a month or quarter). This includes the sale of produce, meat, dairy or other farm products, as well as products that are resold, such as jam in a farm store, or produce brought in from other farms. It is colloquially referred to as the "top-line," as it is what you bring into the business before noting any expenses.

   This can be broken down further into broad categories. You may choose to break out the revenue by sales channel—for example, what you earned at the farmers markets, wholesale, CSA, and farm store. Alternatively, you may break out the revenue by product, such as eggs, meat, produce and flowers. My suggestion is that you chose one or the other. If you try to track sales channel at the same time as product, things will get messy (if you sold eggs at the farmers market, do you track it as farmers market sales or egg sales?). If you're using QuickBooks, you can do both, and we'll cover that in chapter 6, Setting up QuickBooks.

   Note: Grants are not included in the top-line revenue/sales. They appear in the "other income" section (see #7 below), or on the statement of cash flows.

2. **Cost of goods sold (COGS):** cost of the items that have been resold. This is the direct expense of the items you sold. If you have a farm store, it is the actual cost of the produce and other products that you purchase to resell. Some farms choose to include the cost to grow and sell produce (such as seeds, labor and packaging) in COGS, but I prefer to include only the cost paid by the business to buy products for resale.

TABLE 2.1. Sample Income Statement

Revenue	
Crop Sales	148,662
Total Revenue	148,662
Cost of Sales	21,768
Gross Profit	**126,894**
Direct Operating Expenses	
Booth Fees	2,155
Equipment	1,983
Fertilizer and Lime	1,045
Mulch	5,050
Pest Control	1,381
Seeds and Plants	3,594
Supplies	8,889
Total Direct Operating	24,097
Payroll	
Labor Hire	42,283
Taxes: Payroll	
Worker's Comp + Disability	1,871
Total Payroll	44,154
General and Administrative	
Accounting Services	265
Advertising	1,081
Bank Fees	440
Insurance (excluding health)	749
Continuing Education	30
Meals and Entertainment	164
Office Supplies	772
Permits and Licenses	10
Professional Fees	450
Total General and Administrative	3,961
Repairs and Maintenance	
Car and Truck	3,847
Gasoline	4,837
Repairs and Maintenance	292
Tools	1,347
Total Repairs and Maintenance	10,323
Occupancy	
Rent or Lease: Other	3,325
Utilities	3,000
Total Occupancy	6,325
Total Operating Expenses	88,860
Operating Income	38,034
Interest	664
Depreciation	12,549
Income Before Taxes	24,821
Taxes	652
Net Income	24,169

For many farms, COGS is minimal, as they often limit reselling others' products. For farm stores, this number is important because it allows business owners to garner further insights about the business (see gross margin below).

3. **Gross profit (margin):** sales minus COGS. This is the amount of money available to cover expenses and still have something left for profit. Gross profit refers to the dollar amount (sales minus COGS). Gross margin refers to the percentage (gross profit/sales).

   The gross profit is a really important number for several reasons. First of all, it tells you what you have left after purchasing your goods for resale to cover operating expenses—paying your crew, rent, office supplies—and hopefully still have a profit. You also want to take a look at your gross margin to evaluate how much you are paying for the goods you resell and how much you charge your customers. If your gross margin is too low, it could be one of four things:
   - You paid too much for your inventory.
   - You didn't charge enough to your customers.
   - Someone stole from you.
   - You threw away product.[3]

   For a production farm operation, this number is not as important as it is for a farm store that purchases goods for resale.

4. **Operating expenses:** the expenses associated with producing and selling your product, as well as generally managing the operations of your business. We will further break down the operating expenses into five categories (see below).

5. **Operating income:** gross profit minus operating expenses. A positive number tells you your core operations are profitable and by how much. A negative number indicates the core operations of the business are not profitable—that is, expenses exceeded revenue.

   Some people may refer to this as the "bottom line," but they are misusing the term. When I hear people talk about the bottom line, I always try to clarify if they are in fact talking about the operating income.

6. **"The Line:"** an imaginary line separating the operating income and expenses from non-operating income and expenses. Often, there are revenues and expenses that you want to see on the income statement, which are not directly related to the operations of the business, such as rental income or interest expense. They are still recorded on the income

statement, but separated so you can have an unobstructed view of your operations.

7. **Other revenue and expenses:** can include items that are tangentially related to the company's general operating activities. These include interest income or expense, as well as grants, rental income, or proceeds from the sale of equipment. Off-farm income really shouldn't be on your income statement, but if you really want to include it, you would do so here. These items are "below the line."

8. **Income before taxes:** operating income less other revenues and expenses.

9. **Income tax expense:** taxes on income.

10. **Net income:** income before taxes less income taxes. This is your "bottom line."

For tax reporting purposes, using the above categories to organize your revenue and expenses is sufficient. I've seen income statement expense categories aligned with the Schedule F[4] or in alphabetical order. Organizing the operating expenses further into categories, however, makes it easier to gain insight into the business performance. Here are the major categories:

a. **Direct operating expenses:** this includes seeds, soil amendments, animal feed, mulch, packaging…anything that is directly tied to the operation of your business.

b. **Labor:** this includes payroll (salary and hourly), payroll taxes and employee benefits, such as health insurance or worker's compensation, and contract labor.

c. **General and administrative:** these are the tangential expenses associated with running your business, such as advertising, office supplies, insurance, telephone, internet, web hosting, and bank fees.

d. **Repairs and maintenance:** in theory, this is also a direct operating expense, but I break it out separately because it's a prime farm expense and worth tracking separately.

e. **Occupancy:** this includes rent, property tax and utilities: the cost to be on the land.

For a complete list of suggested account categories for a farm business's income statement, see Appendix 1. You may opt to add or delete categories based on your business.

Keeping It Clean

When I met Austin, he had been farming part-time for 20 years, while maintaining an off-farm day job, and was in the midst of planning his "retirement" to farm full-time. Austin had developed a strong reputation in the community for selling high-quality beef and poultry to schools and institutions. With Austin planning to rely on the farm for his primary income, we wanted to make sure he had a profitable business. Honestly, he wasn't interested in working with me. As he said, "I don't have time for the minutiae of the financials." He was too busy marketing and selling, and was confident that his business manager was running things just fine, thankyouverymuch.

Nonetheless, I poked around his books.[5] Total Annual Revenue for 2012 was $1,009,171; and Net Income was $242,304. At first blush, that seemed pretty good! But as I looked deeper into the numbers, I saw that grant income was mixed in with sales income. In fact, the grant income totaled over $700,000 for the year! Granted (pun intended), some of the expenses were related to the grants, but when they were all taken out, the farm had actually lost close to $200,000 for the year.

By separating operating revenue and expenses from the other revenue and expenses we could see that the business on its own was not profitable, and relied on grants to maintain viability. This small adjustment to the income statement made a huge difference in the story of the farm and informed decisions of how it wanted to operate.

What's not included on the income statement is the purchase, or sale, of assets. An asset in the example of a farm store could be the shelving units, refrigeration, or large displays. It could be large equipment such as tractors, and structures like a barn or greenhouse. These assets that you purchase are not actually going to be on the income statement: they appear on the balance sheet. We'll talk about that next. Also not on the income statement is money that you borrowed to purchase equipment. Similarly, if you took out a loan, and received cash, it is not recorded as revenue because borrowing money is not what you do throughout the natural course of running your business. Loans will be on your balance sheet and on your cash flow statement.

The Balance Sheet—What You Have and How You Got It

Also see this video for more details: juliashanks.com/the-balance-sheet/

When to Use

- When applying for a bank loan or grant, lenders and grantors will want to see a balance sheet. They will want to see what assets you have and the debt-to-equity ratio.
- When managing cash flow, the balance sheet can give you a heads up on what cash will come in the door soon, and what will be owed soon.
- To understand the changes in your business, you can compare one year's balance sheet to previous years.

Key Features

- A snapshot in time, usually at the end of the fiscal year or month that details what the business has (assets) and how it got them. Did the business borrow money (liability)? Or did it earn it (equity)?
- It shows who has claims to your assets—whether it is you/your business or someone else such as a vendor, bank or investor.

The balance sheet lists your assets, liabilities and equity. It is called the balance sheet for two reasons:

- It's the ending balance of all your asset (A), liability (L) and equity (OE) accounts.
- It's always in balance: assets always equal liability plus owner's equity. The balance sheet equation can be represented as A = L + OE. Or OE = A – L.

The Balance Sheet Equation

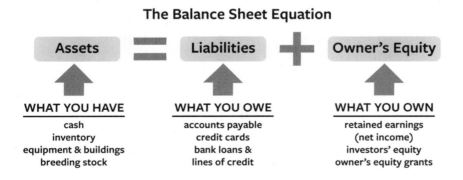

Assets	**Liabilities**	**Owner's Equity**
WHAT YOU HAVE	**WHAT YOU OWE**	**WHAT YOU OWN**
cash	accounts payable	retained earnings
inventory	credit cards	(net income)
equipment & buildings	bank loans &	investors' equity
breeding stock	lines of credit	owner's equity grants

The balance sheet is one of those things that most entrepreneurs don't give a hoot about, but lenders always want to see.[6] If you plan to apply for a loan or get a grant, you'll need to be familiar with it, and produce, at the very least, a rudimentary one.

Assets—What You Have

The academic definition of an asset is something that has future economic value as the result of a past transaction.[7] Simply put, assets are things that your business possesses. Assets include cash, savings, money your customers owe you (accounts receivable), vehicles, equipment, and inventory. The past transaction can be as simple as an equipment purchase. It could also be that you sold produce for cash. The future economic benefit is that the asset is cash, or can be used to generate cash.

Liabilities—What You Owe

The textbook definition of a liability is something that has a future obligation as the result of a past transaction. Simply put, liabilities are things that your business owes. Liabilities include money that you owe your vendors (accounts payable), a bank loan or a credit card balance. Liabilities don't necessarily align perfectly with assets. As an example, you may have an outstanding loan from a bank for $10,000. You may have purchased a greenhouse with it, purchased seeds, and maybe a little is left over as cash/working capital in your bank account. Similarly, the cash in your bank account doesn't align with other assets or liabilities: it can be "acquired" through sales of your product or a loan.

Owner's Equity—What You Own

The equity in your company comes from three places—money that you personally invest, money that others invest in your company (which could be in the form of an investment or a grant), and retained earnings. Owner's Equity represents the porportional value of assets that are owned by you—the business owner.

Retained earnings are the cumulative profits the business has generated. Most often, owners plow profits back into the business, though some of the profits may be saved as cash in the bank. Retained earnings do *not* directly correlate to money in the bank or money you paid yourself. As an example, let's say, last year was Solstice Farm's first year of business; and it earned

$10,000 in profits (revenue minus expenses). Farmer Dan may have cash in the bank or used the earnings to purchase equipment; either way, he now has an asset which he owns. Remember the balance sheet equation? Assets = Liabilities + Owner's Equity; what you have, and how you got it. Retained earnings show that you, the owner, have a claim to the equivalent value of assets. If Farmer Dan purchased a greenhouse with the earnings, he owns the greenhouse because he purchased it with retained earnings.

If Dan took the profits and used it for personal expenses, then he would note that as "Owner's Draw." He has taken equity out of the business, and that has decreased his retained earnings. In other words, the earnings were not retained in the company.

In the first year of business, retained earnings equals net income. In the second year of business, retained earnings = year one's retained earnings + year two's net income (or revenue minus expenses, less any owner's draw). Revenue and expenses directly impact retained earnings, and therefore, owner's equity.

Formatting the Balance Sheet

Within each broad category (assets, liabilities and owner's equity), there can be many accounts—inventory, loan balances, vehicles, etc. Assets are listed in order of how quickly they can be converted to cash. Liabilities are listed in order of how soon they need to be paid. An asset whose total value can be realized within one year is considered "current." A liability that is due within one year is considered current. As an example, credit card debt is due to be repaid every month, therefore it is considered current. A car loan that has a five-year term is considered a long-term liability. Inventory typically is sold within a year, and is therefore considered current. By contrast, a tractor is a long-term asset because it gives value for many years.

The Statement of Cash Flows—Where Cash Is King

Also see this video for another explanation: juliashanks.com/cash-flow-statement/

When to Use

- For projecting ebbs and flows in cash so you can better plan.
- To understand where the cash is coming into and out of your business.

Key Features

- Divided into three sections: cash flow from operations, cash flow from financing and cash flow from investing.
- It is for a range of dates, such as a month, quarter or year.
- It represents changes (flow) of cash, whether it's into the business (cash inflow) or out of the business (cash outflow).

It doesn't matter how much you sell, or how profitable your business is on paper, if you don't have the cash, you will be out of business. As they say, "Cash is King!"

The cash flow statement summarizes the sources and uses of funds over a given period of time. It can explain fluctuations in the cash balance, particularly when it's out of sync with the general operations. For example, you may have a profitable year, but you're running low on cash. This could be explained from having made a purchase in equipment or paying off a loan. Similarly, you may have plenty of cash in the bank despite operating at a loss. You may have financed your operations with credit cards or another kind of debt.

Six Hands Farm

Michael had successfully managed other people's farms for two decades and was finally ready to strike out on his own. He understood the ins and outs of crop planning and farmers market sales. He could manage the payroll and schedule employees efficiently. He wrote a business plan and created financial projections.

He found a prime piece of farmland, just outside of Asheville, NC, and secured a mortgage to purchase it.

His first year on his own land had its usual ups and downs, but nothing extraordinary. Michael survived and felt pretty good. In fact, when he looked at his P&L, he posted $13,000 in profit! But his bank account told a different story: he checked his bank balances, and there was barely $1,000 left. He had fallen behind with his equipment supplier and now owed close to $5,000. What happened?

Two things happened. First, when he did his financial projections, he did not account for debt service. That is, his monthly bank loan payments of $2,000 were not factored into his projections. Even though he made $13,000 in profit, that wasn't enough to pay his loan, so he fell behind on his other bills. Second, not all his customers paid on time, compounding cash flow problems. So while he had the sales, he did not always have the cash in his pockets (or bank account).

Certainly, he knew how to run the business, but he failed to take into consideration the impact of cash flow and debt on his core operations. With this newfound understanding of his cash flow, he realized he needed to set higher sales and profit targets in order to better manage his debt.

When thinking about cash, there's cash flow and cash position (i.e., cash balance). Cash flows in and out of your business on a daily basis. You can have a negative cash flow (more cash left the business than came in) while still having a positive cash position (money in the bank).

Net Income ≠ Cash Flow ≠ Cash Position

Cash Flow from Operations (CFO)

This section summarizes the net increase (or decrease) in cash as a result of operating your business. It starts from a baseline assumption that the net income is the total cash flow from operation (which we know isn't true, as Michael's story demonstrates), and then adjustments are made.

Some of the adjustments:

- **Accounts receivable** (also see Accrual Accounting, page 24). You may have recorded sales from your customers, but not yet received the cash. This is particularly relevant if you're selling wholesale, and some customers take as long as 30 days to pay. If you have a balance in accounts receivable, you received less cash than your revenue suggested.
- **Accounts payable** (also see Accrual Accounting, page 24). You may have purchased supplies but not yet paid for them. If you have accounts payable, then there is more cash available than the net income would suggest.
- **Unearned revenue.** Your CSA customers pay in advance, but the money is not earned until you deliver produce. You have the cash, but not the

TABLE 2.2. Example of a statement of cash flows.

Cash Flow From Operations	
Net Income	1,920
Adj for Depreciation	6,214
Total CFO	**8,134**
Cash Flow From Investing	
Mower	2,500
Box Truck	15,000
Greenhouse	8,000
Total CFI	**(25,500)**
Cash Flow From Financing	
Inflows	
Credit Card Loan	25,000
Grant 1	2,000
Grant 2	1,000
Loan 1	25,000
Grant 3	9,250
Total Inflow	**62,250**
Outflows	
Loan Principal Repayment	3,182
Credit Card Repayment	25,000
Total Outflow	**28,182**
Total CFF	**34,068**
Total Cash Flow	**16,703**
Beginning Cash	5,000
Ending Cash	21,703

revenue. And when the produce is delivered, you have the revenue/sales, but do not receive any cash.

- **Depreciation.** This non-cash expense gets recorded on the income statement, and it's good to leave it there for a whole host of reasons, but it needs to be added back for cash flow. (For more on depreciation, see page 28.)

Cash Flow from Investing (CFI)

Purchases in equipment, land, buildings and infrastructure are considered investments in the business—and the cash used for these purchase is recorded in the investing section as a cash outflow. Similarly, if you sell equipment, the cash you receive is recorded as a cash inflow.

Cash Flow from Financing (CFF)

Inflows and outflows from cash that relate to loans, grants or equity investments are recorded in the CFF section. Only the principal portion of loan repayment is recorded on the cash flow statement; the interest is recorded on the income statement.

This cash flow statement shows that the investments in the business (the mower, box truck, and greenhouse) were purchased with funds from a loan and a grant. The credit card financed a short-term cash need, as witnessed by the quick repayment in full.

Julia's First Law of Accounting and Newton's Third Law of Physics

For every action there is an equal and opposite reaction. This is Newton's third law of physics and the underlying concept of "double entry accounting." We can think of every transaction as having two components. For example, if you purchase seeds and get an invoice for it, you get the inventory (first component), and then you have accounts payable (second component). Another example: If you sell product, you receive cash (first component) and your owner's equity goes up from the revenue (second component).

Action	Equal/Opposite Reaction
Take a loan	• Cash increases (money received from loan) • Liability Increases (loan balance increases)
Purchase a tractor	• Equipment assets increase • Cash decreases
Sell product	• Revenue increases (retained earnings increase) • Cash (or A/R) increases
Pay down debt	• Cash decreases • Debt/liability decreases

Tying It All Together

The three financial statements operate in alignment; a change in one statement affects the other two.

Incorporating the concepts in Julia's first law of accounting and the balance sheet equation $(A = L + OE)$, let's lay out a few transactions to illustrate, a *very* simplified view of a business's operation.

	Assets		Liabilities	Owner's Equity		
				Owner's	Retained Earnings	
Transactions	Cash	Tractor	Loan	Capital	Revenue	Expense
1. Beginning Balance	0	0	0	0	0	0
2. Open Business	10,000		10,000			
3. Buy Tractor	–5,000	5,000				
4. Seeds/supplies	–2,000					–2,000
5. Sell Produce	8,000				8,000	
6. Ending Balance	11,000	5,000	10,000		8,000	–2,000

1. Before you open your business, you may have a twinkle in your eye, but no assets, liabilities or equity. The starting balances in all accounts are zero.
2. In order to start your business, you borrow $10,000. You now have a loan payable, as well as the $10,000 cash in your bank account.
3. With the proceeds of the loan, you purchase a used tractor for $5,000. Your cash balance decreases by $5,000 and you now have an asset worth $5,000: the tractor.

4. You also purchase seeds. This is an operating expense, so not only does your cash balance decrease, but so does your retained earnings.
5. You sell produce at the market for $8,000 (obviously, we're disregarding a whole host of expenses here for the sake of keeping this example really simple). You now have an increase in cash as well as retained earnings (resultant from the sales revenue).
6. The ending cash balance is $8,000. You have a $5,000 asset (the tractor), a $10,000 loan (liability) and $6,000 in retained earnings ($8,000 less $2,000), which is owner's equity.

The numbers in the cash column populate the statement of cash flows. The numbers in the retained earnings column feed into the income statement. And the ending balance is your balance sheet. Ultimately, each transaction impacts all three financial statements.

| | Assets | | Liabilities | Owner's | Retained Earnings | |
Transactions	Cash	Tractor	Loan	Capital	Revenue	Expense
1. Beginning Balance	0	0	0	0	0	0
2. Open Business	10,000		10,000			
3. Buy Tractor	−5,000	5,000				
4. Seeds/supplies	−2,000					−2,000
5. Sell Produce	8,000				8,000	
6. Ending Balance	11,000	5,000	10,000		8,000	−2,000

Statement of Cash Flow · Income Statement · Owner's Equity · Balance Sheet

Underlying Principles

For a video tutorial on Accrual vs. Cash Accounting, see these videos: juliashanks.com/cash-vs-accrual-accounting/ and juliashanks.com/revenue-recognition/

Cash vs. Accrual Accounting

Cash and accrual are two methods for tracking revenue and expenses.

Cash accounting: Revenue is recorded when you receive the cash, and expenses recorded when the cash leaves your bank account. This means that if you have a CSA, you're recording revenue in January through March when your customers are prepaying for their subscription. But the expenses of harvesting, packaging and delivery won't occur until July. Cash accounting is good for understanding the cash flow of your business. But it can be difficult to understand other metrics, such as the cost of labor relative to your sales.

Accrual accounting: Revenue is recorded when the product is delivered (and the revenue is earned) or the expense is incurred. Many times, the cash transaction does not occur at the same time as the revenue being earned or the expense being incurred. As a result, with accrual accounting, you will have accounts receivable (A/R) and accounts payable (A/P).

- **Accounts Receivable:** If you record revenue but don't have the cash, then there's a balance in the A/R account. This can happen if you make a wholesale delivery, but your customer does not pay for 30 days. You have the sale, but not the cash. A/R is noted on the balance sheet as an asset; it's a future benefit (cash payment) as the result of a past transaction (delivering produce).
- **Accounts Payable:** Similarly, if you record an expense, but haven't paid it yet, there's a balance in the A/P account. If, for example, you purchase packaging supplies, but the vendor doesn't require payment for 15 days, then you incurred the expense, but haven't paid the cash yet. A/P is noted on the balance sheet as a liability; it's a future obligation as the result of a past transaction.

Accrual is better for understanding the profitability of the business, as you can match revenues with expenses incurred to earn the revenue.

Using the example of the CSA, you can better understand the benefits of accrual accounting. All the money comes in during January through March when you sell your subscriptions. With cash accounting, the income statement shows a wildly profitable month. If you look at your bank account, you're thinking "I'm rich, I'm rich!" but in fact the money isn't yours yet. You will have several months of delivering product (May–October) when you don't have money coming in. So May through October will look highly unprofitable despite working your tail off. You need to recognize that even though you'll have sales in the late summer, you won't have commensurate cash coming as a result. Further, if you want to make sure your labor costs are on track, it will be easier to understand the effect of payroll as a percentage of sales.

Let's say you're running a flower business. All your customers order flowers in December for the holidays, but some of your customers put it on account and don't pay you until January. If you use accrual accounting, you will show all the revenue in December. If you use cash accounting, you will show revenue when the cash was received; some in December and some in January.

Implications
- With accrual systems, you will use accounts receivable and accounts payable.
- When you look back at your business next year, and want to plan, cash accounting shows you the seasonal flow of cash; accrual accounting allows you to understand the seasonal flow of business.
- With accrual, you can more easily align your expenses with your revenues and evaluate if they make sense.

Just for the record, I prefer accrual accounting. I find it easier to look at expenses relative to revenue and make decisions. It's also easier to see what money is coming down the pike (accounts receivable) and what money is due very soon (accounts payable).

Everything Has a Bucket (or a shoebox or a folder)

Back before the ubiquity of computers, many farmers and entrepreneurs saved their receipts in shoeboxes or folders. They had a folder for each category of transaction: a folder for fuel receipts, a folder for hardware store receipts, another folder for sales receipts from customers, and so on. All transactions[8] had a place. The same is true with modern accounting. Every transaction has a place—whether it's accounted for on the income statement, balance sheet, or statement of cash flows, or as an expense, income, asset, liability or equity. Further, every transaction can be categorized by the type of expense, income, asset, liability or equity.

Implications
- If you properly track all revenues, expenses and other cash flows in the appropriate category, you will have a better understanding of your business and how to make adjustments.

Materiality

Some things just aren't worth the effort, like trying to keep the dust out of the driver's console of your truck. And the same can be true in accounting. Is it worth tracking red plastic as a separate category if you only spend $100 a year on it? Probably not. It's easier to glom together "plastic" into one category, and perhaps even combine it into the broader category of mulch. To be sure, I'm not suggesting you *don't* track the $100. But I don't think it's worth having its own category.

Implication

As you set up your bookkeeping and tracking systems, there's a balance between granularity of detail and materiality.

Variable vs. Fixed Expenses

How do your expenses change as your business grows? Do they stay the same regardless of sales, or do they increase?

Some expenses stay flat—like the fee to your accountant to file your taxes, your cell phone bill, or the annual web-hosting charge. Whether you have $100,000 in sales or $500,000 in sales, the annual expense will stay constant. These are fixed expenses.

Other expenses change depending on your sales. As sales increase, you hire more staff to work the fields and manage the farmers market stand. In order to grow sales, you grow more product and raise more animals, and therefore purchase more seed, soil amendments, and feed. These are variable expenses because they vary (and increase) as your business grows. Variable expenses are usually represented as a percentage of sales. That is, what percentage of your sales do you spend on variable expenses?

Implications

- Understanding which expenses are variable and which are fixed will help you when creating financial projections or enterprise budgets.

Common-sizing Numbers

If you looked at your income statement and saw that you spent $10,000 on seeds last year, would that seem like a lot? It depends on what your sales were. If your sales were $50,000, yeah, that's a lot! If your sales were

$100,000, that would be okay. If your sales were $150,000, I'd congratulate you on running a tight ship. And if your sales were $500,000, I would question why your seed expense was so low (or whether you saved seeds).

It's hard to look at numbers without context. The best way to give numbers context is to "common-size" them, that is, to look at the numbers as a percentage of sales. This allows you to compare your expenses to other years. Did you spend more on seeds this year than last year? The best way to judge is to look at the number as a percentage. Further, you can compare your numbers to industry standards. If the farmer down the street spent 5% on seeds and you spent 7%, you can consider why that happened. Does he get seeds from a different source? Is she more efficient? Is she purchasing conventional when you purchase organic? Certainly, there are many explanations, but now you have a data point to explore the health of your business.

Depreciation and Accumulated Depreciation

For more on depreciation, see this video: juliashanks.com/depreciation/

Depreciation is a way to allocate the value and usefulness of an asset over its life. It also serves as a reminder of what you need to save in order to replace your assets when they expire.

Let's say, for example, you buy a used tractor for $5,000, and you expect the tractor to run for another five years. Every year that you own the tractor, it gives the farm value. And every year, the dollar value of the tractor diminishes. Simply stated,[9] you use $1,000 in value of the tractor every year. The value the tractor provided for your operations is tracked on your income statement as an expense called depreciation. Each year, for five years, you will record $1,000 in depreciation expense for the tractor.

After the first year, the tractor will have depreciated $1,000, and its value will, therefore, be $4,000.

After the second year, the tractor will have depreciated another $1,000 for a total of $2,000, which is the accumulated depreciation. Accumulated depreciation is the running total of how much value an asset has lost over its usage. Accumulated depreciation is tracked on the balance sheet, decreasing the value of the assets you have.

The IRS publishes depreciation schedules.[10] These tell you how to depreciate assets for tax purposes.

Depreciation and Accumulated Depreciation

DEPRECIATION EXPENSE

Example: You buy a tractor for $5,000 on January 1.
You anticipate it has a 5-year life, and at the end of year 5, it has no value.

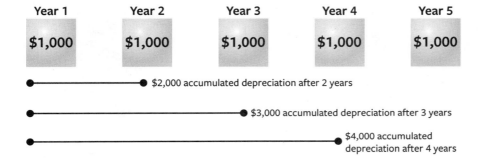

Where Do We Go from Here?

The financial statements are the endgame…so how do we get there? Where do we find the numbers to create them?

If you're creating historical statements—a summary of your business operations—you'll get them from past transactions, which you can track in QuickBooks. If you're creating projections about future performance, you will need to create assumptions.

Creating projections: Go to chapter 4: The Business Planning Process.
Creating historicals: Go to chapter 6: Setting up QuickBooks.

Notes

1. It is also referred to as the profit and loss, or P&L. You will see these terms used interchangeably.
2. You may hear people refer to things "above the line" or "below the line." They are talking about revenue or expenses that are related to either the operations of the business (above the line) or secondary operations (below the line). The line is drawn at operating income.
3. If you buy milk to resell at your farm store and it expires before you can sell it, you may throw it away (feed it to the chickens or take it home for your personal use).
4. Schedule Fs are the attachment to your tax returns that are designed for farmers to report farm income and expenses.
5. Austin was required to work with me because of a grant he received.

6. For more details about what investors look for in the balance sheet, see Chapter 5, Financing Your New Venture.
7. Virginia Soybel and Robert Turner, Babson College, Massachusetts.
8. A transaction is any event that is recorded in the financial statements.
9. Using straight-line depreciation, which means that you are evenly allocating the cost of the asset over its useful life.
10. You can find the depreciation schedules here: irs.gov/pub/irs-pdf/p946.pdf

CHAPTER 3

Planning
Your New Venture

Most people dream of running their own business: following their passion, being their own boss and having control of their time. It sounds glorious!

The reality of entrepreneurship (and farming) is far less glamourous. In addition to the actual farming you love, you need to market, sell and deliver your product, do bookkeeping, manage seed inventory and harvest schedules, repair and maintain equipment, and so on. And if you don't do all these things yourself, then you also hire staff, delegate and manage. If you spend 50% of your time actually farming, then consider yourself lucky!

Before investing your life savings or taking a loan you can or cannot afford, ask yourself a few questions to make sure you're ready for entrepreneurship. Because, let's face it, when you start your own business, you are no longer a practitioner (farmer), you're a business owner and entrepreneur.

The overarching question is: Do you believe you have all the skills, energy, money, people and knowledge to start a business?[1] The following questions are not meant to deter you, rather to ensure that you go into this with your eyes wide open to the realities of entrepreneurship.

Deciding to Take the Plunge

Becky was always curious. She had spent the last 10 years working on other farms, always asking questions of her bosses. "Why are you doing that? Why are you doing this?" She went to workshops and seminars, and took courses through Annie's Project. After all that experience, Becky had too many of her own ideas; she was ready to venture out on her own. As she said, "I'm independently minded, stubborn and driven. I'm the type of person who gets bored after too many years working for others."

In her first two years of business ownership, she realized she had to be a generalist. Of course, she had to master organic farming techniques. But she was also head marketer and book-keeper. And she drove the delivery truck into the city every week. "I didn't think about all the challenges until I was doing it myself," she said.

Nonetheless, she can't imagine doing anything else. "I wanted the real life experience of why food has value. And that comes from ownership and self-sufficiency of running my own business."

Skills and Knowledge

1. **How much experience do you have farming?** Can you make a crop plan? Do you know what types of soil you need for the crops you want to grow? If you don't, consider getting an apprenticeship.[2]
2. **How strong are your business skills?** Can you manage people and money? Do you like working with people and selling?

People

3. If you don't have all the skills you need, **what resources can you access to round out the business?** Do you need a business partner who understands the business side of the operation while you focus on the farming? Can you connect with your local agriculture agency?[3] Do you have an accountant to help with bookkeeping and taxes? What other resources can you tap?

Energy

4. **What's your shit tolerance?** As they say, "Everything sucks some of the time."[4] This couldn't be more true for farming and entrepreneurship. Can you stand doing things you don't like for 50% of your day?
5. **Are you willing to work as many as 90 hours a week?** Unlike many in-

dustries, there's immediacy to farming. The irrigation needs to be set up today; it can't wait until tomorrow or your crops will wilt. The cucumbers need to be harvested today, or they will be too big and bitter. The chickens need to be fed today; they can't wait until the weekend, when you're catching up on chores.

On top of that, all sorts of unexpected things will not go according to plan. I can't tell you what they will be, but I can assure you it will happen. All your plans for an organized day, with a balance of work and life, will seem like fantasyland after a few months of running your own businesses.

All of this adds up to many more hours a week of working than you anticipate, especially in the first years of business while you're getting your systems in place. To be sure, you will not work at this pace 52 weeks a year, but it can be the norm and not the exception.

How hard are you willing to work?

Money

6. **Can you afford to not earn money for the first 6 to 12 months of business?** As the owner of a business, you are the last person paid. Your vendors, suppliers and employees need to be paid before you can take a draw. During the lean times, you may not be able to pay yourself at all. Do you have enough savings? Do you have a spouse who can support you? To that end, if you're applying for a loan, the bankers will want to know that you are not putting yourself in financial jeopardy in taking the loan. If you get evicted from your apartment because you've fallen behind on rent, then there's an increased chance that you will fall behind on your loan payments, too.

   In general terms, it's always a good idea to have 6 to 8 months of savings in case you lose a job. The same is true for starting your own business. Make sure you have enough savings to carry you through the lean times. Chances are you won't have the time during the growing season for a second job if you're short on cash.

7. **What is your tolerance for risk and uncertainty?** In addition to bolstering your savings to support yourself during the start-up phase, you will likely need to borrow money to start your own farm, whether it's to purchase land, equipment or a greenhouse. Some people don't like

borrowing money, and that's okay. It means you need to plan and save more. If you use your own money, are you okay losing everything you put into the businesses? I'm not suggesting that you will, but it's always a possibility. And as they say, plan for the worst and hope for the best.

If you decide to borrow money, the business planning process will help you figure out what you can afford to borrow and what you need to borrow to launch your business. Often, it can be a loan amount that exceeds anything you've borrowed before, and it can feel scary. Are you comfortable borrowing money? And are you willing to put your credit at risk if things don't work out?

You're feeling bold and confident that you can make it work! Awesome! Let's get started.

Next up: What's Your BIG (or small) Idea?

Clarifying Your Vision

If you don't know where you are going, any road will get you there.

LEWIS CARROLL

Having the desire and will to start a business is most of what you need to get started. And, of course, you need an idea. It doesn't have to be a big, bodacious change-the-world idea, but you need something to define the destination.

You need to state your *desire* (what you want to do), clarify your *driving* passion, and *define* success.

Desire: In three sentences or less, state what you want to do. Here are some examples:

- I want to start an organic vegetable farm that grows Asian vegetables and herbs. I will sell through farmers markets and CSAs.
- I want to create a full-diet CSA by growing vegetables and grains and partnering with other farmers in the community.
- I want to build a meat processing facility so that I can process my own animals. I would like to expand my business to process animals for other local meat producers.

_____

_____

_____

_____

_____

Drive: When things get tough, what's going to keep you moving? What's that burning force inside that will keep you going when times get rough? This can be a mission statement or just a goal. Here are some examples:

- I want to create a positive impact on the local food system.
- I want to prove that we can eat locally—completely and year-round.
- I want to ensure that meat producers have the infrastructure they need.
- I want my children to have a better place to live than the one I grew up in.
- I want to ensure that no one has to eat foods with pesticide.

What is your driving passion?

_____

_____

_____

_____

_____

Define: You know what you want to do; how will you know you have achieved success? Here are some examples:

- I'm earning $50,000 a year in salary.
- I can support myself on the farm income and take vacations.
- We will have built up the health of our soils and pastures in order to grow abundant food for our community.
- Our farm will be supported by a dedicated customer base of local people who feel connected to the farm in a meaningful way.
- We will have created year-round employment for others who wish to learn and farm with us.
- We will have established a financially sustainable farm business that provides our family with a stable income and rewarding work.

What is your definition of success?

_____

_____

_____

_____

_____

Farming and entrepreneurship are much too hard to go into haphazardly. Make sure you're setting yourself up for success.

Getting Started Checklist

- I answered the 7 questions and decided I have what it takes.
- I can state what I want to do in 3 sentences or less.
- I clarified my driving passion and mission.
- I have a definition of success.

Notes

1. inc.com/ss/6-questions-ask-starting-business
2. Here are some options for farming internships in North America, mostly the US: https://attra.ncat.org/attra-pub/internships/
3. For a list of state agencies in the US, go here: rma.usda.gov/other/stateag.html. In Canada, you can start here: agr.gc.ca/eng/home/?id=1395690825741 or farmstart.ca
4. markmanson.net/life-purpose#.oo78dg:w1cW

CHAPTER 4

The Business Planning Process

Thinking About Your New Venture and All Its Possibilities

When I first got on the phone with Farmer Rebecca, she was so excited and nervous about her plans. Breathlessly, she told me about her ideas to sell fresh-pressed juices made with the produce she was already growing on her farm. And start a farm store to sell the juices, along with her produce. And maybe purchase land to increase production. She wasn't sure if she needed to borrow money and wasn't quite sure about the process for launching her business. She had just too many unanswered questions: what sort of permit did she need? How much should she charge for her juices? Does a farm store make sense? Can she afford to purchase/rent land in the right location for the store?

Her enthusiasm was infectious, but I knew that getting her ideas on paper would help to clarify her vision, and how all the different components fit together. Working together, I could help her decide if the idea was feasible. How much juice would she need to sell to make this a worthwhile venture? Did she have the customer base in her small town to support a farm store?

Getting Started

Writing a business plan and creating financial projections can help clarify your vision, evaluate the feasibility of your plan and, if you need financing, prepare you to approach investors and lenders.

For your own planning, you'll need two components:
- A business plan
- Financial projections

These two items go hand-in-hand. The business plan is the story behind the numbers. And the numbers detail how you will make the plan work.

If you are approaching investors, you will also need an executive summary, a one-page document that provides an overview of the business plan, and financial projections (see juliashanks.com/TheFarmersOfficeTemplates/).

The Business Planning Process

The business plan and financial projections are created in tandem; and the process is anything but linear. As you flesh out your concept in the written document, the numbers will evolve in your projections. And as you modify the numbers to create a viable business model, the narrative will adjust to reflect the economic realities of your business.

The Business Planning Process

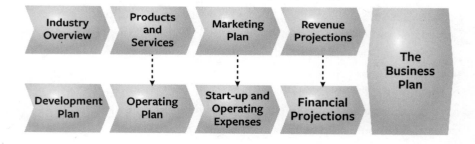

The industry overview provides insights into where there are gaps in what consumers want, and what products and services would fill the space. After you decide on the products and services you want to provide, you can then lay out your operating plan (how you will actually run your business to create the products and bring them to market). Knowing what you want to sell within the context of your industry guides the marketing plan. And the marketing plan gives you a sense of how much you can sell.

With the operating plan, you can map out both your start-up expenses and operating expenses. With revenue projections, start-up expenses and operating expenses, you now have the information needed to create financial projections.

Of course, if something changes, whether it's the products you want to sell or your operating plan, you will need to revisit your financial projections.

Continuation of Rebecca's Story

The flurry of ideas overwhelmed Rebecca, and she decided to focus first on the juices before expanding further. In her original thinking, she would sell pressed juices at the weekly farmers markets. Her additional overhead costs would be low because she was already at the markets selling vegetables. She would charge $5 per juice, and her costs would be $1 per juice for the ingredients and $.20 for the cup, leaving her with a nice $3.80 profit per serving.

She pondered how much she could sell at each market. If she sold 50 portions, she would profit $190 per market, and $6,840 for the 18-week season, attending two markets a week. Good, not great.

She thought through the sales process further. If she pressed the juices to order at the market, she would bring an extra staff person with her to work the press and sell the juices. The "theater" of pressing juices to order could certainly increase sales, but would cost another $75 per shift. If she premade big batches of juices, she wouldn't require the extra staff at the market, but would need someone to make the juices in a rented kitchen, cost-ing about $300 per month for the kitchen and $100 a week for the labor. The "good, not great" above was becoming less good, and certainly not good enough to purchase land.

Whether Rebecca set her goals on purchasing new land or not, she realized, she needed to expand her selling opportunities, and started exploring wholesale channels.

With each run of the numbers, Rebecca revised the story she was telling. Originally, she was going to sell juices at the farmers market, and expand the amount of land she grew on. When the numbers didn't add up, she considered new sales opportunities. And with that came new labor and overhead costs. As the numbers changed, she added new details to the story. The back-and-forth between the numbers and the story yielded a more cohesive plan.

Rebecca decided the best way to test the feasibility of the new venture was to just start. She found a used juicer for just a few hundred dollars, acquired her Food Handler license for $50, and was able to start selling at the farmers market with minimal investment. This allowed her to better understand the expenses involved in making and selling juices, as well as the interest from her customers.

Even if you complete a business plan with financial projections before you start the new venture, the numbers and story will further shift as you gain experience and have actual numbers to track. The business plan is a living document that will adjust as your plans grow and evolve.

The level of detail in your plan and numbers depends on how much money you need and who you're asking. When Rebecca is ready to purchase a higher-capacity $5,000 juicing machine, she can approach friends and family, and won't need too much documentation, at most a quick budget and summary of her plan (level 1). As the venture expands (purchasing land and expanding her sales channels), she will need more money from banks and government grant agencies. They will want more details (level 2). If Rebecca instead purchased land to expand farming operations and build a farm stand, her investors would require even more documentation (level 3).

What Do You Need?

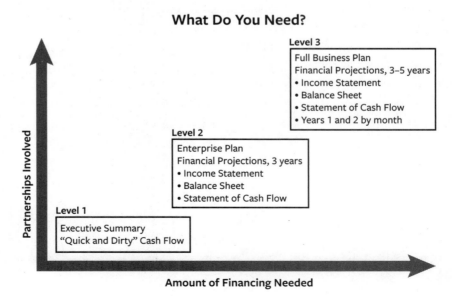

Level 3
Full Business Plan
Financial Projections, 3–5 years
• Income Statement
• Balance Sheet
• Statement of Cash Flow
• Years 1 and 2 by month

Level 2
Enterprise Plan
Financial Projections, 3 years
• Income Statement
• Balance Sheet
• Statement of Cash Flow

Level 1
Executive Summary
"Quick and Dirty" Cash Flow

Partnerships Involved

Amount of Financing Needed

Level One

- "Quick and Dirty" cash flow statement—2 years of income statement projections with capital purchases and debt service detailed
- An executive summary that highlights the growth and operations strategy, investment required and returns expected

Level Two

- An enterprise plan, which is a pared-down business plan that emphasizes the growth and operations strategy, with limited analysis of the industry, markets, and critical risks
- Financial projections for three years, including an income statement, balance sheet and statement of cash flows
- Detailed assumptions behind the financial projections
- A list of the sources and uses of funds

Level Three

- A complete business plan, including industry analysis, marketing plan, and critical risks
- Financial projections (income statement, balance sheet, and statement of cash flows) for five years, with the first two years broken down by month
- Detailed assumptions
- Sources and uses of funds
- Executive summary

The more money you need to finance your business, the more numbers investors will want to see. More money equals more at risk for the investor. They will take more time with their due diligence to make sure it's a safe investment.

Business Plan Writing Guide

Most farmers and entrepreneurs view the business plan writing process as a necessary evil to getting a loan or a grant. Many business owners consider the plan obsolete after it's written. Paul Brown explains in *Forbes Magazine*: "…you can plan and research all you want, but the first time you encounter something you didn't expect out in the marketplace, the plan goes out the window. Once you are underway, things never go exactly the way you anticipate."[1]

So true. Things rarely go as planned. But that doesn't mean you shouldn't write a business plan. The process of writing a business plan forces you to think through all aspects of your business. Do you have a market for your product? Do you know how you will manage your operations? Do you know what equipment you need to purchase? Do you know how you will find customers? Do you know the order in which you'll make things happen?

You can answer all of these questions (and more) through the process of writing a plan. It certainly won't provide all the answers to solve the problems that come up along the way; but it ensures that you've started thinking about them and gives you the resources to troubleshoot them more quickly.

Further, the business plan can provide a compass for when you get off-track. After the launch of your business, you'll likely be mired down in the details: planting, weeding, feeding, selling and so on. It's easy to lose sight of where you want to go. The plan is written as a viable business model, and provides a benchmark of what your business needs to do each month. If you see your revenues falling short of projections, or expenses exceeding expectations, then you can make adjustments before it's too late. You may need to modify where and how you sell to increase revenue, or scale back expenses if they are out of line with revenues.

The best way to get the full benefit of the process is to follow an outline or template. This ensures you don't skip over key sections. For a complete guide and outline on how to write a farm business plan, see Appendix 2.

Creating Financial Projections

Financial projections are really just a feasibility study. Will your new venture provide enough profit to pay back loans and give you, the business owner, and your investors greater value[2] as a result? Is it worth the time, effort and financial investment to launch or grow your business? In other words, what is the feasibility and practicality of this new venture?

You can answer these questions by creating financial projections.

In tandem with testing the financial feasibility through projections, also consider the market feasibility. That is, are there enough customers who will buy your product at the price you want to sell it? Certainly you can create financial projections that demonstrate wild success, selling silage at

$100/pound. But if you don't have a customer base willing to buy it, then the plan isn't feasible.

The *Income Statement* tells you if the venture will be profitable.

The *Balance Sheet* lets you know if you're building equity in your business. For the financier,[3] it shows how much debt you are taking as a ratio to your equity. The greater ratio of debt to equity suggests a greater risk in the venture.

The *Statement of Cash Flows* clarifies that you can afford to take on the financial risk of borrowing money and investing in land, equipment and infrastructure, and maintain a positive cash position.

The financial projections also communicate the story of how you'll succeed to potential investors, lenders and grant agencies. They have three main objectives:

- To force discipline and objectivity into your business dream through a methodical approach.
- To demonstrate a thorough understanding of your company's business model, building credibility with your potential investors.
- To have the tools to answer the question "What if?" What if sales are lower than expected? What if you purchase a greenhouse? What if…

How do you create these statements, and where do the numbers come from? They come from assumptions you list about your business:

- What will it cost to launch and grow your business?
- How much financing will you need?
- How much money will you earn in revenue?
- What will it cost to run your business?

As one farmer said to me, "If you can't put numbers to it, don't even bother. You're wasting your time and money."

———————————— **A Note about Templates** ————————————

There are hundreds, if not thousands, of Excel templates around the Web to help farmers create financial projections. I have yet to see one that really works, mine included. The challenge with these templates is that they are difficult to customize. The formulas and cell references zip around the worksheet and multiple tabs. If you make an adjustment to suit your specific business, links break, and then the template becomes dysfunctional. My suggestion: create the various components of the projections (assumptions, income statement, balance sheet, cash flow and debt service) independently, and then manually insert cell references to link them together.

It can take several weeks to put these components together. It's not easy; and few people besides the number geeks (like me) enjoy it. Most entrepreneurs (farmers as well as chefs, food producers, techies and so on) rely on consultants to help them.

That said, whether or not you create projections yourself, you should understand the process. First, it simplifies the work with the consultant or technical assistance provider. Second, you'll recognize the various checkpoints where you can evaluate the feasibility of the numbers. You don't want to go through the whole process and discover you need to go back to the beginning because the numbers don't add up.

The complete process of creating the financial statements is detailed in Appendix 4. What follows is an overview.

1. Start with Historicals (to create assumptions)

"Performance data quoted represents past performance; past performance does not guarantee future results."

Investment firms offer this standard disclaimer to warn investors about potential risks. Despite the risks, past performance is still the best indication of future results. When creating projections about future business performance, the best starting place is the past. What did your business do last year? What did you sell, what were your expenses, how much was payroll, and so on. The future numbers will, no doubt, differ; but you have a baseline to begin.

When launching a new venture, you don't have historical numbers to pull from. But numbers exist; others have forged paths before you to start farm businesses, so there is precedence. If you currently work on a farm, ask your boss about some of her numbers. Or find a mentor who has run a similar business to get a better understanding. We'll go into more detail in the next section.

I often encourage new entrepreneurs to just start, like Farmer Rebecca, in whatever capacity they can. If you can find a quarter-acre to farm on with a low cost to start, do it! It will give you a better sense of all the numbers and little details you need to think about.

2. List Operating Assumptions

Whether or not you have historical numbers, you'll need to list assumptions about how you think future operations will go. How much will you sell? How many people will you hire? What will your expenses be? If you're starting with a new venture without historical numbers, it can feel like smoke and mirrors with pixie dust sprinkled on top. Nonetheless, you will have to make educated guesses until you have real numbers; then you can revise your projections.

Forecast Sales

There are two ways to create a sales forecast: start with your revenue goals (top-down) or start with specific production goals (bottom-up). If you have historical sales information about your business, you can use these as a starting point: estimate how you want the business to grow and evolve in the coming years.

Top-down Sales Projections

For a farm that has been in business for a while, it can be more efficient to start with the total amount that you want to sell (or earn) in a year or how much you want to grow from current sales; let's say it's 11% growth from $90,000 to $100,000. From there, break the number down further—how much money do you anticipate earning from different revenue streams: from, say, attending farmers markets or selling wholesale?

If you decide that 75% of your revenue will come from farmers markets and 25% from wholesale, then you need to earn $75,000 from the farmers markets and $25,000 from wholesale.

- The typical farmers market runs for 22 weeks, so that's about $3,500 in targeted revenue per week.

- If you can earn $1,750 on the average market day, then you'll need to attend two markets per week. If your average market yields $1,200 per day, then you will need to attend three markets.
- The growing season is about 26–30 weeks. That translates to about $900 per week in wholesale.
- If the average wholesale customer orders $200 worth of produce, then you need four or five customers.

Bottom-up Sales Projections

Especially in the first year of a new business, it's easier to start from the bottom up with sales projections.

- How many wholesale customers do you want? How much do you expect them to order each week? How long is your season?
- For how many CSA subscribers do you think you can grow produce for?
- How many dairy cows will you have? How many gallons of milk can they produce in a day? What is your annual yield?

For example, if you think you can grow enough produce for 40 CSA members, and your price will be $500 per share, then your estimated revenue from the CSA would be $20,000.

Both bottom-up projections and top-down projections are based on averages. You won't sell $30 worth of produce every day to 25 customers in your farm store. Some days will be more, some less. You won't sell $1,750 worth of produce at every farmers market. In the early spring months, it will be less, and in September, it will be more.

Gut Check

For a new or beginning farmer, with limited sales experience, assessing the likelihood of achieving these numbers can be challenging, and the best you can do is a gut check. After you've determined how much you will earn from each revenue stream and how the numbers will break down, sit with it. How does it *feel*? Do the numbers feel doable or totally audacious?

If the sales projections seem optimistic, even if doable, scale them back. There's no shame in conservative numbers. Better to create a business model that works in less than ideal circumstances than one that relies on achieving aggressive goals to succeed.

Now that you have a good understanding of your sales goals, look back at

Gut Check

Aaron from High Spirit Farm had been pounding the pavement, talking with chefs and grocers about buying his produce. It was early March when we sat down, and he was pretty sure that he had five commitments for purchasing throughout the season; he felt confident he could sell $45,000 worth of produce.

We took a top-down approach to understanding the numbers. "Bob, the chef over at Foley Restaurant, said that I was the only local farmer he was buying from. I bet I could sell $15,000 worth of produce to him"

"Really?" I asked. "That's almost $700 a week. How many cases of produce is that? 14?"

"Yeah, that's about right," he responded.

Aaron called Bob to better understand his commitment. In the beginning of the season, when all Aaron was growing were greens and radishes, 14 cases would be ambitious. At the height of the season, 14 cases seemed about right.

With the insight from his customer, Aaron could adjust his sales projections. Either he would need to find more wholesale customers to sell to in the spring in order to meet his projections, or he would need to scale back his sales estimates.

the marketing section of your business plan. Do you have a strategy to find the customers to make the sales a reality?

Estimate Expenses

If you have historical numbers:

- Review each line of your historical financials.
- Look at the history to see the patterns or trends. Do some expenses increase as the sales volume increases? These variable expenses (including cost of goods sold) will be estimated as a percentage of revenue. Home in on a percentage of sales that you will use to estimate the projected expense. For fixed numbers, such as accounting services or insurance, estimate your expected expenses for coming years based on what you spent in previous years.

If you are launching a new venture:
- Review the list of expense on the sample income statement (Appendix 1). Skip over labor for now. We will address that separately below.
  - Which of these expenses do you anticipate having?
  - Are there expenses not suggested that you think you'll have?
  - For each expense, estimate what you'll spend each month or year. Where possible, research and get quotes: how much does feed cost, and how much do you think you need? Call an insurance broker and find out what kind of insurance you'll need and how much it will cost. Call a payroll company to understand payroll expenses. How much do black plastic, seed trays and soil amendments cost? How will you package your product, and what does it cost? It's tempting to make up numbers, what you *think* will happen, but there's no substitute for actual numbers and real research.
  - Consider how often you will pay for these expenses. Some things, like seed purchases or liability insurance, may be paid once or twice a year. Others happen regularly, once a month, like your rent, utilities or phone bill.

─────────────── **Miscellaneous Expenses** ───────────────

It's tempting to add a catch-all category, like "miscellaneous expenses" (with a projected cost of, let's say, 5% of revenue) to cover all the niggling little things you may have forgotten in your projections. While there's nothing wrong with bumping up your projected expenses to plan for the unexpected, this shouldn't be used as a means to gloss over the little expenses. These "little" expenses can be things like a cash box for the market or labels for your packaging. Twenty dollars here, twenty dollars there; all these little things add up to real money, and can be the difference between a successful venture and a fledgling venture. Bottom line: don't use a catch-all category.

───

Estimate Labor Cost

One of your biggest expenses will be payroll. If you set a budget for $10,000, how many employees can you have? Is that enough to support your sales goals? Will you pay payroll taxes and worker's comp insurance? As your business grows, your payroll expense will also grow.

Creating a sample schedule provides a framework for estimating labor expense.

- List out the different positions you need filled—harvesters, CSA managers, farmers market crew, drivers, etc. You don't need to know *who's* going to be working, just that *someone* will be working.
- Set the hourly wage for each position.
- How many people do you need working each shift, each day?
- How many hours will they work in a day? How many days per week?

Once you determine the number of shifts worked per week, the duration of each shift and the hourly pay, you can then calculate an estimated weekly payroll expense.

In addition to the wages you pay to your employees, you will also pay federal and state taxes, social security and worker's comp insurance on behalf of your employees. This can vary from 9% to 12% in additional expense.[4]

For simplicity sake when creating financial projections, add 11% to the salaries for payroll withholdings and insurance. For actual day-to-day operations, I encourage you to seek a payroll service provider to ensure that you are properly withholding payroll taxes and submitting them to the appropriate agencies.

As your business grows, presumably so will your payroll. As you raise more chickens, milk more cows, harvest more produce, you will need a larger crew. The payroll calculation above can be used as the baseline for future year by calculating the payroll number as a percentage of sales.

If, for example, you anticipate wages to be $27,000 in your first year and your revenue to be $130,000, then wages as a percentage of sales is 20.7%. In subsequent years, you can use 20.7% to estimate your payroll expense, or slightly less if you expect to gain efficiencies.

3. Calculate Your Projected Operating Profit

With the above revenue and expense assumptions, you can create a profit and loss statement that brings you to the operating profit. For a formatting guide, see Chapter 2 for a review of operating profit, and for an Income Statement Template, see juliashanks.com/TheFarmersOfficeTemplates/

Entrepreneurial Thinking

A few more expenses will be factored into the income statement after the operating income is calculated (interest expenses, taxes, and depreciation), but this is a good place to stop and evaluate the venture and projections. At its core, does your business model make sense and cents? Some questions to ask:

- Can you support yourself from the operating profits and/or pay yourself a salary?
- Are the profits sufficient to grow the business? Should you decide to build a new chicken coop or milking parlor, can you afford to save money to finance it?
- If the business experiences a hiccup (the tractor needs an unexpected repair), will you have enough breathing room financially?

As an entrepreneur, consider whether the business generates enough profit to satisfy your financial needs: grow your business, pay your debt and withstand the unexpected expense. Also consider that you may crave the occasional luxury (new Carhartts or a dinner out) and a vacation every once in a while. Sure, you love what you do, but that elusive "work-life balance" is important too.

Further, the operating profit also begins to give an indication of how much debt financing the business can afford. The monthly loan payments certainly cannot exceed the operating profits, or you will run into cash flow problems; better still, profits after paying the debt service should still leave breathing room for the above.

If the business is not as profitable as you think it should be at this stage of the projections, take a break and revisit your assumptions. Profits will only shrink further as debt service and taxes factor in. In addition, things never go as expected—though you carefully planned out your assumptions, something will be off. You won't get into a farmers market, your first planting of tomatoes will get flooded by heavy rains, or a predator will get into the chicken coop. The projections need enough wiggle room to absorb the inevitable setbacks.

4. Investments in Your Business

You're creating the business plan and projections presumably because you want to make some sort of financial investment in your business. List the planned improvements and purchases you want to make and their expected cost. With the list of improvements, you have an idea of what your depreciation expense will be, and also what sort of financing you need.

Do you need to borrow money or get a grant, or can you finance the purchase with earnings from the business? If you borrow money, then you will pay interest on the loan as well as repaying the principal. Depending on the stage of planning you're in, you may or may not know for what type of loan you qualify. At the very least, call a few banks to get interest rates and loan terms. Use the Debt Service Template at juliashanks.com/TheFarmers OfficeTemplates/ to calculate the monthly payments. For more details on financing options, see Chapter 5: Financing Your New Venture.

5. Calculate Net Income

With an understanding of your investments, you can calculate depreciation expense. With an understanding of your financing needs, you can estimate interest expense.

You now have all the numbers necessary to complete the income statement projections. From the operating profit, you will add other income and subtract other expenses (depreciation and interest). From the subtotal (EBT, Earnings Before Taxes), you can estimate taxes. This brings you to the net income.

To summarize how we got to net income:

Revenue – COGS = Gross Profit

Gross Profit – Operating Expenses = Operating Profit

Operating Profit ± Other Income and Expenses = Earnings Before Taxes

Earnings Before Taxes – Taxes = Net Income

Entrepreneurial Thinking

While it is not uncommon to have net losses in the first year or two, the business should turn a profit by year three. If the business does not project profits within a time frame that feels comfortable to you, then revisit your assumptions.

It is possible to have a positive operating profit and negative net income. This results from interest payments that exceed what the business can afford and/or depreciation expense. As you continue on to create the balance sheet and cash flow projections, pay close attention to the cash balance to make sure it stays positive.

6. Create the Balance Sheet

Balance sheet projections start with the beginning balance sheet and then layer in the expected changes in the assets, liabilities, and owner's equity. Unlike the income statement process, this is more about just plugging in numbers.

The Beginning Balance Sheet

You can also use the Balance Sheet Template. For a refresher on the balance sheet, review Chapter 2: Building the Foundation.

Whether creating a beginning balance sheet for a new business or for a current business, the process is the same. For a new business, you will want to project what you expect to have on the day you start your business (after you have made the initial purchase of equipment and received financing). List your assets and liabilities; and then calculate owner's equity by subtracting the value of your liabilities from the value of your assets.

For the purposes of business planning, a "business" balance sheet is sufficient. You do not need to list personal assets and liabilities. However, some banks and lenders will also want to see a "personal balance sheet," that is, what do you own and what do you owe that is not part of the business? This could include your personal vehicle, your house, the mortgage on your house and/or student loans.

Balance Sheet Projections

Balance sheet projections should be created in the same time interval as the income statement projections. If you project out income for each month, then your balance sheet projections should also be by month.

The work you've completed to this point provides you most of the numbers you need. You know the assets you have or will purchase; you know the debt you will take on; you know the profits you expect to earn. The one thing you don't know is how much cash you will have, but it's a simple calculation since you have all the other numbers. As you may recall:

$$\text{Total Assets} = \text{Total Liabilities} + \text{Owner's Equity}$$

Written another way:

$$\text{Cash} = (\text{Total Liabilities} + \text{Owner's Equity}) - \text{All Other Assets}$$

And if we turn this equation around, we can solve for cash:

Cash = (Total Liabilities + Owner's Equity) – All Other Assets

Solving for cash provides an estimate for your cash balance.

The balance sheet projections outline what you anticipate your assets and liabilities will be at the end of each period (whether it's a year or a month), and also portend how your cash balances will be affected.

Entrepreneurial Thinking

As you review your balance sheet projections, look at the cash balance for each month. Does it dip below zero? Mathematically, negative cash is possible, but in reality, it is not. If your cash balance is negative, then someone is not getting paid and checks will bounce. This results in fees and penalties, unexpected loans or credit card debt.

If the cash balance goes negative and then recovers (that is, you show a positive cash balance in later months), the issue can be the timing of purchases and debt payments. You may need to delay some expenses until your cash position improves. If the cash balance dips below zero but does not return to a positive balance, then there are core issues in your assumptions.

Revisit your work so far. So many assumptions and decisions have been made to get to this point, and many ways in which things went astray that led to a potential of negative cash. Certainly, this isn't a foregone conclusion—it may turn out that your actual revenues are higher than you project and your expenses are lower than you expect. Nonetheless, you should troubleshoot a negative cash balance and figure out ways to avoid it:

- **Is the net profit high enough to support all the purchases you want to make and debt you take on?** To that end, are your expenses too high? Are your revenues too low? What changes do you need to make?
- **Did you appropriately capitalize your business?** That is, did you borrow the right amount of money? Did you borrow enough money to carry you through start-up? Did you borrow too much such that you cannot afford the monthly payments? Are the terms of your loan prohibitively expensive?
- **Does your equipment purchase schedule align with your cash flow?** Do you plan to purchase more equipment than you can afford? Do you need to delay some purchases? What changes can you make to ensure that you project a positive cash balance at all times?

7. Create a Statement of Cash Flows

For a refresher on the Statement of Cash Flows, refer to Chapter 2. A Cash Flow Projections Template can be found online (see juliashanks.com/The FarmersOfficeTemplates/).

The balance sheet tells you (among other things) if you've run out of cash or not. The cash flow statement provides more detail.

It is divided into three categories:
- Cash Flow from Operations (CFO)
- Cash Flow from Investing (CFI)
- Cash Flow from Financing (CFF)

The most common mistake that entrepreneurs make when creating cash flow projections is that they ignore the investing and financing components—and essentially recreate the income statement with a slightly modified form. As you know by now, there's more to your business than just the operations; you have financing and investing activities.

Cash Flow from Operations

CFO equals net income plus/minus any non-cash transactions. These transactions include:
- Depreciation: While depreciation appears on the income statement as an expense, it's not an actual cash flow. The cash outflow associated with the purchase happened months, if not years, ago, and was shown in the cash flow from investing section.
- Accounts Payable: If you track your expenses on an accrual basis, you may have expenses for which you have not yet paid. You may have charged them to your credit card or received terms from your vendors. The expense appears on your income statement, but the cash did not flow out of your business (yet!).
- Accounts Receivable: If you track your revenue on an accrual basis, you may have sales for which you have not yet received cash money.

For the purposes of these cash flow projections, we will ignore Sales Tax Payable. Depending on where you live, you may collect sales (or meals) tax for some of the goods that you sell. You receive the money from your customers and then pass it along to your State revenue department. This money is not revenue, and it does not belong to you, even though it can be in your bank account for as long as three months. If you properly account for sales tax, it will not show up on your income statement, but it will on the balance sheet.

Cash Flow from Investing

All the capital purchases that you made and recorded on the balance sheet are noted in the CFI section of the cash flow statement. Even if you received financing to make the purchases (which would be a cash *inflow* from financing), you still need to note the *outflow* from investing.

If you sell equipment, then the money received will also be noted in this section as an *inflow*.

Cash Flow from Financing

Money comes in and out of your business through financing. If you receive money from a lender or investor, this is an *inflow*. If you pay down debt (such as a car payment, mortgage or operating loan), this in an *outflow*.

By noting the cash flow from financing, potential investors can see if you financed your investing purchase through your operating revenues or through loans. If your business remains cash positive despite continued operating losses, the cash flow statement will show where the money is coming from. These details you probably know intuitively, but an investor will want to see on paper.

8. Sensitivity Analysis

Throughout the process of building these projections, you've stopped to make sure everything's on track and the numbers add up. Presumably, at this point, you're confident the numbers work, and you've written a plan for a viable business.

Nonetheless, it's worth scaling back your projections one more time. Just in case. Just in case your start-up expenses are more than you expect. Just in case it takes longer to build your customer base. Just in case any number of things don't go as planned.

Over the years, I've worked with many clients. Something always goes wrong. Here's a list of some of things that happened:

- The grant they hoped to get was not awarded.
- The building contractor disappeared. After a new one was found, he disappeared too.
- It rained too much.
- It didn't rain enough.
- The growing fields had more boulders in them than expected.
- The primary growing field flooded during heavy rains.
- Rodents burrowed holes in the fields.
- The town didn't let the farmer put a sign on the highway directing potential customers to the farm stand.
- The cider press took longer to restore than anticipated, and a full apple season was missed.
- The zoning board wouldn't allow a kitchen build-out without an investment of $10,000 more for additional infrastructure.

For your own peace of mind, and that of your investors, test out the worst-case scenario. Go through your assumptions:
- Scale back your revenue assumptions by 15%
- Increase your start-up costs by 10%
- Increase your operating expenses by 10%

What happens to your profitability and cash flow? Do you need to borrow more money? Can you afford to borrow more money? Play around with the numbers to find the worst case where you can still build a viable business.

9. Revisit Your Business Plan

With all the back and forth with the numbers, the story of how your business will work may have changed. Go back through your business plan and make sure the narrative matches your numbers.

You put a lot of hard work into these numbers. Much of this process is for your benefit—to ensure that you launch a viable business. But the reality is that you went through this process because you want to approach lenders and investors to help you finance your business.

If you seek investors or bank loans, the financiers will want to see that the business is sufficiently profitable such that you can pay back the loan and afford to live a reasonable lifestyle.[5] There's more to pitching investors than just handing them your business plan and financials; and for all this effort, you want to put your best foot forward.

You have a business plan, you have financial projections; you're ready to start approaching investors and lenders.

Notes

1. forbes.com/sites/actiontrumpseverything/2013/08/14/why-business-plans-are-a-waste-of-time/
2. Value in this instance can be defined as money, as increased land value, increased local food available, or whatever is important to you.
3. The person or institution giving you money to finance the venture—it could be a bank, grant agency or family member.
4. Both employees and employers pay a portion of the payroll taxes. The employees' portion is deducted from their paycheck.
5. If you burn out because you work too hard for too little money, then the investor loses her money.

CHAPTER 5

Financing Your New Venture

"It takes money to make money."

A new greenhouse costs upwards of $8,000. Increasing your herd size can cost $1,200 per head. For sure, in time, these investments will pay for themselves—depending on the size of the greenhouse, it can generate as much as $20,000 a year in revenue; and cows will generate $6,750 each in revenue after they're processed. This assumes you've already acquired (and financed) land to produce on; if not, then buying land may be necessary. These are just two examples of purchases that help you grow your business. But where do you find the money to buy them?

You have four options:
- Save money
- Borrow from friends and family (F&F)
- Look to outside organizations
- Get prepayment from your customers and fans.

Save Money

If you're uncomfortable borrowing money, and many people are, you can finance your growth with your savings. These savings can come from your personal bank account, or by saving money that you earn through your farm business. Expanding your operations by purchasing only what you can afford outright means slower growth with less risk than the other options. There is no shame in slow and steady growth.

That said, sometimes it's worth spending the extra money (and taking a loan) to buy the appropriate equipment, infrastructure or land. An extra $5,000 to buy a better tractor can save you money (from costly repairs) and time (when a lesser tractor may be out of commission for repairs).

Borrow from Friends and Family

Friends and family (F&F) are often the first source of start-up financing; it's a tempting option. The loan process is less onerous, the interest rates are generally more favorable and the payment schedule more forgiving. If you fall behind on payments, family is less likely to impose fines and penalties. It's the epitome of "patient capital": friends and family are patient in the repayment of their investment.

In ideal circumstances, you clearly define the terms and pay back the loan exactly as planned. But troubles ensue if expectations are not managed.

Unfortunately, F&F loans tend to be less formal with the terms, and skip over the necessary conversations about contingencies and what-ifs. "It's family, after all," you might say. "We'll just work it out." But. If for some reason you miss a payment or otherwise fall behind, you (the entrepreneur) may just assume your family will let it slide. You may assume that it's really an interest-free loan, when it isn't. If the lender, your friend or family, has a different expectation than you, then difficulties will arise.

Just remember—you won't see your banker at Thanksgiving and weddings—you will see your F&F. Unless your family is okay being patient (and potentially turning the loan into a gift), then you should consider other options for financing.

Outside Organizations

Most often, entrepreneurs get financing from outside organizations and individuals.[1] Financing can come in the form of debt, equity, grants and crowdfunding. No matter how you raise the funds, there is a cost to it. As they say, nothing in life is free.

The different financing options vary by the terms of repayment: what sort of interest/return is provided to the financier; and when the funds are expected to be repaid. As a general rule, the higher the risk for the financier, the greater returns they expect.

Debt

Debt is borrowed money that is repaid to the lender in a specified time period with a pre-determined interest rate, and is usually in the form of credit cards or loans.

When entrepreneurs pay for goods with their credit card, they owe the bank the money within a certain number of days (usually around 30 days); money that is not repaid by that date accrues interest and penalty fees. It is quite easy to access money through credit cards, and the banks do not enforce rules on how or when you repay the money; it benefits them when you miss a payment. This creates a high risk for the credit card company, for which they charge a high interest rate (anywhere from 15%–23%). *I strongly discourage using a credit card to finance your business.* Interest rates are exorbitant, and if you fall behind on a payment, the debt can quickly get out of control. For more thoughts on that, see Chapter 9: Stabilizing the Business.

Lines of credit are similar to credit cards in that the bank gives you a borrowing limit for a short-term loan, typically one to three months. They are typically used for operating expenses during periods of slow cash flow. Interest rates are low for the first few months, but then increase as time passes. With a line of credit, you only take what you need.

Loans also require repayment with interest of the money borrowed, but the repayment schedule is more structured. Lenders define the terms within a certain time frame (from 1 to 30 years) with a set interest rate. The repayment schedule is fixed, and payments are usually expected every month (though some only require a single annual payment). Banks tend to be more conservative, and require the borrower to fill out an application and submit a business plan to demonstrate that they can repay the loan in a timely fashion. Banks further minimize their risk by taking collateral and restricting who they lend to. As such, they can charge a lower interest rate than credit cards. In addition to banks, non-profits and for-profit organizations, as well as individual investors, provide loans. Many of these "non-bank" lenders have a mission to support sustainable agriculture and are willing to take a greater risk: lending money to otherwise "unbankable" entrepreneurs at affordable rates.

Equity

Equity investors provide entrepreneurs money to grow their business in exchange for a partial ownership stake. The amount of ownership they take is relative to the amount of money they provide, as well as the estimated value of the business.[2] The investor then gets a return on her money in two ways: by taking a percentage of the profit distributions, and/or when the business is sold by taking a portion of the sale price (proceeds). They want to ensure your success, as it's the best guarantee that they will get their investment back.

Equity investors come in all shapes and sizes: the patient capital types, which could be individuals, F&F, or institutions; and professional investors who have certain expectations and time frames. It's important to understand who you're dealing with before you sell equity. Equity investors also expect a higher return than a bank; they believe they are taking a greater risk.

Entrepreneurs often misconstrue equity investment as free money because the repayment terms are less rigid than a loan.

Debt vs. Equity

There are many differences between debt and equity. As you decide which option is best for you, consider these differences as they relate to your business and financial needs.

Repayment Structure

Debt is more structured in its payback. This can become a problem if you run into trouble making payments. Banks, especially, are less forgiving about missed or late payments than other types of funders. Non-profit and individual lenders may be willing to work with you. Equity has greater flexibility in repayment to investors, the investor gets repaid when funds are available. If your business is struggling, the investor will forgive slow repayment as they don't want to detrimentally drain your cash flow.

Cost of Capital

While the repayment terms are different between debt and equity, you still pay for the use of the funds in the form of interest or returns. The cost of using someone else's money is typically less with debt/loans than with equity. Whether it's in the form of interest, dividends or returns, you pay for the use of funds.

Ownership Control

The upside of loans is that you do not have to give up management or ownership control of your business. As part owners of your business, equity investors not only have a stake in your business, they may have opinions about how you do things, and can offer unwelcome advice.

Advisors

All financiers want you to succeed, but not all have the resources to provide advisory services to help. Equity investors can offer business advice. Many non-profit lenders offer technical assistance.

Tax Implications

Income tax is based on net income; and interest related to debt is an expense that reduces your income, thereby reducing the amount of taxes owed. Distributions to investors are not considered an expense, and therefore do not impact your taxes. Debt lowers your tax burden.

Planning Future Growth

Some lenders will restrict the number of other loans you take because it increases their risk. When taking out a loan, ask about any covenants that would limit your ability to borrow more money in the future. Often, a growing business will take out multiple loans in order to grow to the next phase.

Spectrum of Success

What will it take for your business to succeed? Perhaps, $100,000 is the minimum revenue you need to break even, and there's a probability that you could earn as much as $300,000, and everything in between. You can barely succeed, moderately succeed, and succeed wildly. That's a broad spectrum. Other businesses either succeed wildly or flop altogether, with nothing in the middle: a binary outcome. The only successful outcome could be if the business is sold to a larger company. This may be true for some value-added food producers or agricultural technology companies.

With these types of speculative businesses, where success is a binary outcome, everyone (both equity investors and lenders) loses their money if the business fails. However, if the business succeeds, equity investors get better returns than the lender. As such, a lender is not likely to offer debt for this type of business, as they don't benefit from the upside.[3]

Prepayment

Advanced payment from customers provides a cash infusion for farmers when they need it most: in the first few months of the growing season when operating costs are high and revenue is low. The most common prepayment system is the CSA (community supported agriculture) model. The customers pay up front for a season's worth of produce. Wholesale distributors will also prepay for goods to help farmers (their suppliers) get a leg up on the season. This benefits the wholesaler as he can secure local product to sell throughout the season.

Prepayment benefits the farmer because it provides cash for the business without the interest payments of traditional financing. But there is a potential downside. When production kicks into high gear, you will not be generating the same volume of *cash* (which is different from revenue) as you would if customers only paid when they received goods. This requires proper cash-flow planning to ensure you don't run into trouble later in the season.

Crowdfunding

In a similar vein to prepayment, some crafty entrepreneurs request donations from friends and supporters. Many websites facilitate this process, including Kickstarter and Indiegogo. This too can feel like "free money" because the money is not repaid. But there are real costs involved in raising money this way. First, the entrepreneur offers some sort of gift to the donor—anything from recognition on a website to a t-shirt to a free CSA share. With the exception of recognition, these all cost money. In addition, you invest time. For crowdfunding campaigns to be successful, you must promote, promote, promote. Tactics include creating and posting videos, regular Facebook postings and email blasts to potential supporters.

The real benefit of crowdfunding isn't the money, but the raised awareness of your business and brand. With young businesses that need to build name recognition, this is an effective way to raise money *and* build loyalty from a customer base.

For an established business with a loyal customer base, crowdfunding can be a source of cash during a crisis. Many farmers have used crowdfunding to raise money when they have lost a greenhouse to a storm or a barn to a fire. In these instances, the farmer does not provide a gift to the donors; it is a way to rally the community.

If your primary goal is raising money to grow your business, you are better off going more traditional routes.

Financing Sources

When entrepreneurs think about financing their business, banks are usually top of mind. Not all entrepreneurs are bankable (for all sorts of reasons from bad credit history to not having collateral). Thankfully, there are many other options, from non-profits and government agencies, to private investors. Below are some options. It is not an exhaustive list as each region in North America has different community organizations, banks and federal agencies.

Banks and Credit Unions

Each institution has its own loan packages and financing options. Contact your regional lenders to explore options that may be suitable for you. Most regions around North America have lenders that focus on the agricultural sector.

Direct Investors

More and more individuals want to put their money where their mouths are, and look for sustainable food ventures to support. Slow Money (slowmoney .org) can help connect entrepreneurs with investors.

Financing Land Acquisition

Farmland is one of the few capital investments you will make where the value typically goes up with time. Unlike leasing, the security of your land tenure will allow you to make better long-term investments and more thoughtful purchases.[4]

Bank mortgages for farmland tend to have higher interest rates than regular home mortgages. Be sure to consult a bank for accurate rates to ensure your business plan supports higher monthly payments than you might expect.

In some regions, land can be prohibitively expensive to purchase. While it's most typical to take on a bank mortgage to buy property, there are increasingly more options available. Companies have sprouted up around North America that provide a variety of solutions. For example, Iroquois Valley Farms purchased over 3,200 acres of farmland, all of which is certified organic or in transition to organic production. They then lease the land back to farmers through long-term tenancies. Similarly, Northeast Farm Access purchases and develops land in the Northeast United States to lease back to farmers.

Many state agencies have programs to help farmers finance land purchases in order to keep the land in agricultural use. These programs pay farmland owners the difference between the "fair market value" and the "agricultural value" in exchange for permanent deed restrictions that preclude any use of the property that will have a negative impact on its agricultural viability.

Government Agencies

Grants and loans are available through a variety of US government organizations, including:

- Farm Service Agency, providing below-market interest rate loans
- State-level departments of agriculture offer grants for land with conservation easements.
- USDA has many grants available through different offices, including:
  - NRCS for greenhouses: nrcs.usda.gov/wps/portal/nrcs/detailfull /national/programs/?cid=stelprdb1046250
  - Value-Add Producer Grants for marketing and advertising
  - Specialty Crop Block Grants.

For more details of federal funding resources, visit the USDA website: afsic .nal.usda.gov/where-can-i-find-agricultural-funding-resources

The Small Business Administration also has a search tool to find loans and grants: sba.gov/loans-and-grants

CDFIs

Community Development Financial Institutions (CDFI) provide financial services (both loans and technical assistance) to underserved markets and populations, including farmers. To find a CDFI in your community, and learn about their specific offerings, you can Google "CDFI" with the name of your region, or search at ofn.org/cdfi-locator

Preparing for Investors

No matter who you approach for financing, whether a bank or equity investor, you'll need to put together some combination of an executive summary, business plan and financial projections (see Chapter 4: The Business Planning Process). These documents help you think through your concept and plans, and they will also demonstrate to potential funders that you've got a good game plan. The bottom line: you're asking someone to give you money. Not only are they investing in your business, they're investing in you. You need to demonstrate that you will be a good steward of their money.

As you prepare your pitch, think about what the investor wants to know to feel comfortable lending to and investing in *you*. You may have heard about the 5 Cs[5]—the five things investors think about when evaluating the opportunity to invest in you:

1. Character
2. Capacity
3. Capital
4. Collateral
5. Conditions

As they look through your plan and financials, and meet with you, they will be thinking about these 5 Cs.

Character

Investors want to see a good business model, *and* they want to see a good business owner. They want to know that you are capable of running your business according to the plan you lay out, and that you know when and how to ask for help. Here are some tips to help you demonstrate your good character.

- **Learn a few words of "Investor-ese."** Every industry has its own language, and when you mix and mingle in other industries, you need to learn a new lingo. This is especially true for farmers applying for a loan or seeking investors to grow their business.

  I recall a conversation I had with an old boss, the president of a food service company. He tossed around the term EBIDA[6] several times. It impressed me that an in-the-trenches chef knew the term—he learned it from years of working with investors.

  Understanding investor lingo will better equip you to answer their questions as they consider investing in you. The Glossary lists and defines key terms that investors and lenders will throw at you.

- **Be professional.** A few years ago, I worked with a farmer who wanted to open up a retail produce store. He showed up to our first meeting in a threadbare t-shirt, tattoos up and down each arm and dropping f-bombs with a particularly ornery disposition. He was an excellent farmer, and I had no doubt that he could build a successful business. I was less confident that he could raise the money he needed because of his personal presentation. In fact, farmers go into their trade because—beyond wanting to make a difference in our local food systems—they don't like offices, and prefer a casual work environment. But if you want to use someone else's money, you need to demonstrate you will be a good steward of their money. And the first step is being professional: from personal appearance, returning phone calls and emails promptly, showing up for meetings on time, and staying positive. In the end, this farmer was able to get the investment needed. In addition to having all the mechanics of his presentation down, he "cleaned up" well, too.

Banks also use credit scores as a measure of your character. If your credit score is questionable, there are ways to improve it. Google "Improve my credit score" for a bevy of resources.

Capacity

Most articles that write about the 5 Cs of credit list "character" as the first C, implying that it's most important. If you ask me, capacity is just as important, if not more so. If your business cannot generate enough revenue (capacity) to pay expenses and repay the loan, then the strength of your character is of little importance.

Lenders and investors will look at various numbers in your financial projections to evaluate the business's capacity to repay its debt, pay the owner and grow the business. The two you should be most familiar with are:

- **Operating Profit:** Does the business generate enough revenue to pay its expenses and pay debt?

   Without doing any great math, operating profit provides a good estimate of cash flow from operations and how much cash is available to repay the debt (both principal and interest payments). You may also hear it referred to as the debt service coverage, and it should be greater than the total debt by at least 20% (for the year).

- **Net Income:** Can the business support the business owner and poise itself for growth?

Remember, depreciation can be used as a measure of how much money the business needs to save in order to continue operations, in essence, the cost to replace the assets as they wear out. Net income incorporates depreciation; a positive net income shows that the business can afford to save money and reinvest in the business.

Many entrepreneurs prefer to pay themselves from the profits, instead of drawing an official salary. The net income will also show what is available for the farmer.

Capital

Capital refers to your financial position and how much equity and cash you have in your business.

In investor-speak, the ratio of liabilities to equity is leverage. Especially for farm businesses, a highly leveraged business (more liabilities than equity) means the business relies on other people's money to grow rather than financing the growth through profits. For an investor, this signals a risky company and investment.

Collateral

Speaking of risk, lenders want to avoid it. One way they minimize risk is by requesting the borrower (you) to put up some form of collateral.

Academically speaking, collateral is an asset pledged as security for a loan. The asset could be your tractor, your home or your vehicle. The lender will look at your balance sheet to see which assets can be used as collateral. If you, the borrower, default on your loan, then the lender can take claim to the assets assigned as collateral. The collateral will be detailed in the loan documents; assets are *not* randomly usurped by investors if you miss payments.

Conditions

Outside forces will impact your business and its success. A drought could hurt your production, and a slow economy may mean fewer people are purchasing organic. Similarly, food safety and nutritional concerns about conventional farmers have been a boon for small farmers. These conditions demonstrate the general business and market climate for your products. Equity investors especially want to see the market demand and growth potential of your business.

Beyond the 5 Cs, you can prepare for investors in other ways.

Tell a Good Story

When asking for money to grow your business, you need to paint a compelling picture. One of my favorite techniques is to start with an introduction like this:

> More and more Americans want to know where their food comes from. They want to eat healthy local foods year-round, but the farmers market schedule is too restrictive and the supermarkets don't offer transparency. *Mother Earth Produce* delivers where other options falter. They provide weekly home delivery of local and organic vegetables, meat and dairy.

In a few short sentences, it states the "customer pain," the opportunity space, and the solution the entrepreneur is providing. It engages the reader quickly.

You will want to weave your story throughout the plan. Consider these questions:

- How does your business enhance the local food system? Will you solve a bottleneck in the system? For example, opening up an abattoir would help local meat producers get their meat processed more locally and easily, with less stress on the animals.
- How will the money help you grow your business? Will you now be able to purchase new equipment to improve efficiencies, or implement a marketing plan to grow revenue?
- How will you be able to pay it back? Can you demonstrate that you will have revenues to cover your operating expenses and also pay back the loan?
- How does your business history portend what your business will do in the future?

Make Sure the Numbers Add Up

Within the words of your story are numbers: increased sales, decreased expenses, increased market share and so on. As you put the numbers on paper (or into an Excel spreadsheet) make sure they reflect your story, and make sense. Make sure that the numbers are consistent and add up. One of the biggest mistakes I see is when entrepreneurs make adjustments to their business plan and forget to update their financial projections (and vice versa).

Be Realistic

Dare I say, be pessimistic. While I generally subscribe to living optimistically, I encourage pessimism when it comes to business planning. Anticipate everything that could go wrong, and plan for it. Recently, I reviewed financial projections for a chef opening a café. She anticipated that she would open within three months of signing the lease and on day one the café would operate at full efficiency and capacity. As it turned out, the café opened four months behind schedule, incurring all sorts of unexpected costs. During opening week, there was a huge blizzard, so they had very few customers.[7] While it's hard to predict everything that could go wrong, it's important to recognize that things will, and to have a plan to address them.

Consider all the bad things that can go wrong in your plan. Can you still be profitable? Investors want to see that in the worst-case scenario you can make it work. Bonus if you achieve your best case.

Meeting investors where they are and making it easy to understand why they should give you their money will go a long way towards helping you secure funding and hit the ground running with your new enterprise.

Dos and Don'ts of Presenting to Investors

- *Do* explain your business concept clearly in one to three sentences. *Don't* run on and on. Readers of your business plan want to be able to quickly capture the essence of your business. The more concise your words, the better.
- *Do* compare your business to others; *don't* ignore the competition. It's important to understand your competition—both as a way to refine your

strategy, and how you will differentiate yourself. It also helps you understand what your customers want.

- *Do* address your strengths; *don't* overlook your weaknesses. When writing a business plan for investors, tell them why they should trust you with their money. The more you can highlight your strengths, the better. At the same time, you need to recognize your weaknesses. When you address your weaknesses, explain how you will overcome them. Further, the more you recognize the gaps in your skills, the better positioned you are to find complementary partners and advisors.

- *Do* share details and information that you think is important. *Don't* expect investors to read your plan. Many investors will read your entire plan, others will skim it, and some will just read the executive summary. When discussing your plan with investors, assume that they haven't read it, and highlight your main points.[8]

- *Don't* text overload; *do* use graphs and charts. Because investors tend to skim business plans, graphs and charts can communicate details about your business and strategy effectively and efficiently. This enables them to absorb more details about your plan.

- *Do* provide enough information so they can intelligently evaluate the opportunity.

- *Don't* assume the reader is an expert in your field (nor wants to be). Depending on who is funding your business, they may or may not know a lot about farming and agriculture. I worked with an entrepreneur to write a business plan for cultivating and selling organic mushrooms: he had 10 pages drafted detailing the methodology and microbiology of cultivated mushrooms. I found it fascinating, but that's the kind of stuff you want to leave out.

- *Do* be succinct in describing your business. *Don't* get mired in the details. The business plan shouldn't detail every aspect of your business, such as what time employees start their shift or uniform is required. You *do* want to provide a summary of the operations and how they translate into costs and efficiencies for the business.

Notes

1. For further reading, see Elizabeth Ü's book, *Raising Dough: The Complete Guide to Financing a Socially Responsible Food Business.*

2. An example: An investor gives you $100,000 to purchase land. The total value of the land with infrastructure is $400,000. The investor then owns 25% of the company.
3. Conversation with Katherine Collins, Honeybee Capital.
4. With the long-term nature of a land purchase, the quality of the soil is most important as it will affect your cash flow for years to come. For more thoughts on acquiring land, see "In the Market for Farmland? Make Soil Your Top Priority by Brett Grohsgal, in *Growing for Market*, February 2004, Vol. 13, no. 2.
5. Depending on where you look, it is also referred to as the 4 Cs, where capital is combined with collateral.
6. Earnings before Interest, Depreciation and Amortization
7. Thankfully, she had enough of a cash buffer to cover the unexpected expenses and lost revenue.
8. You may want to present investors with a PowerPoint deck and/or an Executive Summary. See sample templates at juliashanks.com/TheFarmersOffice Templates/

CHAPTER 6

Setting up QuickBooks

QuickBooks (QB) is the industry-standard accounting software for small businesses, including farms. At a minimum, it's a jacked-up checkbook register. At its best, it is a powerful business tool that provides the information you need to make important decisions about your business.

If you have a basic understanding of accounting, QuickBooks is straightforward, but it definitely takes time to get used to the format and the lingo. Your patience in learning to use it properly will be rewarded.

QuickBooks can help you:
- Balance your checkbook
- Speed up end of year tax preparation
- Understand the profitability of your business
- Track:
  - which customers owe you money
  - which vendors you owe
  - how much time you spend on different tasks and crops
  - which customers and sales channels are most profitable.

For more help on setting up QuickBooks, watch these webinars: Getting Started with QuickBooks and QuickBooks: Expenses and Sales at juliashanks.com/video-tutorials/

Ultimately, QuickBooks will help you track and sort your financial data so you have the information you need to manage cash, develop a growth strategy and create the financial projections that will outline the story of your growth.

Some farmers opt to use Excel spreadsheets to track revenue and expenses. While this method is sufficient for filing your taxes, there's incredible detail about your business that is lost because Excel doesn't have the capacity to sort and analyze the way QB can, and what it can do is not as efficient. As Hannah Wolbach of Skinny Dip Farm says, "I was able to get the information I needed from my Excel spreadsheets, but it took much more labor. With QuickBooks, it's easier to pull the information I need to analyze my finances."

QuickBooks is available in two formats: online (in the cloud) and as a program that you can download and install on your desktop computer. I prefer the desktop version, but both have benefits and drawbacks. If you understand the pros and cons of each, you can decide which option is better for you.

QuickBooks Online vs. Desktop

The Interface

Despite having the same bones, the online system looks very different from the desktop version. If you're used to one or the other, switching can feel quite disruptive. Some of the more basic features of the desktop version seem to be buried in the online version. (For example, if you're trying to find an old transaction, it can be quite laborious.) Some features available on the desktop version aren't available at all online (like mileage tracking). Further, the desktop version lets you easily switch back and forth between different windows; the online version does not allow for more than one window to be open.

Winner: Desktop

System Speed

The online version is a bit slower as every transaction needs to be sent up to the server/cloud and the page reloaded before you can move onto the next transaction. This can be especially frustrating if you have a slow Internet connection (as is often the case in rural communities). As you become more proficient in QB, this slowness will become more pronounced.

Winner: Desktop

Cost

The least expensive online version costs $13 per month. While this is sufficient for the most basic business, I recommend using the "essentials" or "pro" version (at $27 and $40 per month, respectively). These offer more robust reporting by allowing you to sort your revenue and expenses by customer and by sales channel; they also allow you to have multiple users working on the file. QuickBooks offers special discounted pricing for the first 6 months to lure you into this format.

The desktop version costs about $200, and is a fixed, one-time cost. Unless you get a new computer, you don't need to update every year (though QuickBooks will suggest you do). Even if you get a new computer, it's possible to reload an older version. After 6½ months, it's cheaper than the online subscription (or 12 months if you factor in the promotional pricing of the online version).

Winner: Desktop

Collaboration

Depending on the nature of your business and operation, you may want more than one person accessing your QuickBooks files. For example, you may have your CSA manager entering in customer receipts and your bookkeeper entering and paying bills. With the online version, each user can work from their own computer wherever they are, as long as they have an Internet connection.

If you have the desktop version of QuickBooks, collaboration becomes challenging, though there are a few ways to work around this:

- The most straightforward option is to designate one computer in the office for QB and let the different users share it.
- Another workaround, which I've used, is to save the QB data file in the cloud with Qbox or Dropbox and load the software on multiple computers. I can work in QB from my computer, close the file, and then someone else can continue doing other work on the same company file. This method usually works, but sometimes has hiccups.
- The most effective, but also most expensive, option is to host Quick-Books on a server such as Right Network (www.rightnetworks.com/solutions/quickbooks-hosting) or CPA ASP (www.cpaasp.com/). Hosted QB is about the same price as the online version with the other benefits of the desktop interface.

Winner: Draw, leaning towards Online

Technical Support

The time will come when you need help with QuickBooks. Perhaps you'll need to figure out a report or understand why your checking account isn't reconciled. The easiest way for an advisor to help you is to see what's going on in your file. If you have QB online, the advisor can easily log into your account and poke around. If you have the desktop version, you'll need to email the file instead. If you don't have the same version (for example, I have 2014 and you may have 2012), then they can't make any changes without you being required to update your QB ($175). Your advisor can make recommendations, but then you will need to do the corrections yourself. *Winner: Online*

Entering Data on the Go

I was working with one client who would make deliveries, but wasn't always sure what her customers would buy: she would take the original order before delivery, but she sometimes sold an extra case of tomatoes. When she arrived at a customer's shop, she would create the new invoice from her phone using the QB app and email a copy to the customer. Similarly, if she was at the store picking up supplies, she would enter the expense when it happened. Otherwise, she risked losing the receipt in her wallet and not keeping an accurate record.

It's very helpful to be able to create invoices and enter receipts from your smartphone or tablet, especially for farmers who rarely spend time at a desk. With the online version, you can download an app that allows you to manage your books from your desktop or mobile device. *Winner: Online*

While the horsepower and capacity of the desktop version works better for me, you may prefer the convenience and accessibility of the online. I like this about QB: it's flexible enough to handle a wide range of business needs, which allows me to use it with all my clients.

QuickBooks Best Practices

Garbage in, Garbage out

You've heard the expression "Garbage in, garbage out." It means the quality of your end product will be only as good as the ingredients you put in.

If you use mediocre soil, then your tomatoes will be mediocre. If you use high-quality feed, you will raise better-tasting animals. The same is true with QuickBooks. If you don't enter quality data into QuickBooks, you won't be able to extract useful information. With haphazard systems, you'll be lucky if you can file your taxes efficiently. If you take the time to manage QB cleanly, you can glean all sorts of details about your business—what crops are most profitable, what markets are most profitable, if can you afford to purchase that new tractor you've been eyeing. Best of all, tax filing will be a snap.

Consistency

When you set up QB, you'll create a chart of accounts—basically, a list of all the different categories you'll use to sort your income, expenses, assets, liabilities and equity. Every time you make a seed purchase, you'll want to track it as "seed expense." Or maybe you want to track it as "seeds." Or maybe you call them "golden pearls." It doesn't matter, as long as you track it the same way every time. If sometimes you enter them as "seeds" and sometimes you enter them as "farm supplies," then you will have a hard time teasing apart where you're spending your money. Similarly, you will want to track all revenue and expenses the same way. Farmers Market sales should always be entered as Farmers Market sales, and not Produce sales.

Customize for Your Business

When you first set up QB, it will suggest an organizational structure (Chart of Accounts) based on the fact that you are a farmer. But these are *suggestions* for the *average* farmer. A lot of farmers organize their finances according to the Schedule F, but that may not be how you think about your business. You are not bound by the structure QB suggests nor that of the Schedule F.

Create a structure that makes sense for you (within the bounds of general accounting rules). Give accounts names that make sense for you. If you like to call compost "black gold," then that's how you should refer to it in QuickBooks. It will be easier to enter data consistently and garner insight if the structure and account names make sense to you. It's easy enough to translate this into a Schedule F when it comes time to prepare and file taxes.

Make It a Habit

At the end of the day, when you're popping open a beer or pouring a shot of bourbon (or drinking an iced tea), take 10 minutes to empty your wallet of receipts and enter the day's transactions. If you take 10 minutes every day, then it won't become an onerous task. If you wait until the end of the month, or the end of the season, it will be overwhelming.

Whatever You Do, Don't Delete

I've made mistakes, and you will make mistakes, too. We all do, especially when using QB for the first time. Just about all mistakes can be fixed, and QB is quite nimble that way. The only one that can't be undone is deleting a transaction. Deleting transactions can be particularly disruptive if they are linked to others. For example, let's say you create an invoice, receive the payment and then deposit the check. Later, you decide that the invoice was wrong and you want to delete it. Because the check deposit is linked to the invoice, everything will become out of whack as a result. Better to just edit the invoice, than delete and start over.

Note: *The following tutorial for setting up and using QuickBooks uses QuickBooks Pro 2014 for PC as its reference. If you are using a different version or different year, you can find the same features but they may be located in different places. For example, the "preferences" menu is under the QB icon on MAC and under the "Edit" menu on the PC version.*

The Initial Setup

When you first open up QuickBooks, you will want to **Create a New Company** and select **Advanced Setup**. QB will guide you through an interview process. Following are the questions QB asks, with my suggestions on how to answer (and set your preferences). You may change your preferences at any time in the future, by accessing the **Preferences** window.

1. **Enter your company information.** The address and contact information you enter here will also populate invoice and sales receipts templates. Be sure to enter the address as you want it to appear on invoices. You don't need to add your Tax ID number if you don't feel comfortable (QB will use it to generate end-of-year tax forms).
   Click **Next**.

2. **Select Industry.** By knowing your industry, QB recommends tax forms and account settings. Select **Agriculture, Ranching, or Farming**.

3. **How is your company organized?** This will help QB generate reports for tax filings. If you're unsure about your "corporate structure," ask your accountant or select **sole proprietor.**

4. **Select the first month of your fiscal year.** Most often, the fiscal year starts in January. And it's easier to think of the calendar year as the fiscal year. However, if you sell winter and storage crops, such as beets and apples, you grow in the summer and autumn but do not sell out until March. In that case, you may prefer to align your growing cycle with your fiscal year. In this example, your fiscal year would be April–March, and the first month of your fiscal year would be April.

5. **Set up your administrator password.** You can protect your QB account with a password, and no one will be able to access the account without it. This can be especially important if your computer is left in the barn or other public area. If you are unconcerned about password protection, you can skip this.

At this point QB will generate your company file. You will need to save it to your computer. After this, you can start customizing how QB will work for you.

Customizing QuickBooks for Your Business

6. **What do you sell? Services, Products, Both Products and Services.** Do you provide consulting services? Do you ever speak at conferences or bill services by the hour? Do you charge for delivery? These are services. Products are tangible items that you sell such as tomatoes, CSA subscriptions or eggs. **Select Both Products and Services.** Even if you don't sell "services" now, you may in the future; no harm in having this feature turned on.

7. **Do You Charge Sales Tax?** Check your local government agencies if you are unsure about sales tax and/or meals tax on food and other consumables. QB will track sales tax for you.

8. **Do you want to create estimates in QuickBooks?** Some industries create estimates for their customers before doing work, and then convert the estimate into an invoice. Most farmers do not. If you are like most farmers, **select no.**

9. **Do you want to use billing statements?** If you have wholesale clients who do not pay right away, then you will want to track what they owe you. This is especially important if your customers receive several deliveries from you per month. The billing statement allows you to create a summary of unpaid invoices to send to customers. **Select yes.**

10. **Using invoices in QuickBooks.** Do you want to use invoices? If you have wholesale clients, then you will likely invoice them. By using invoices, you will have an easy way to track which customers have paid and which have not. **Select yes.**

11. **Using progressive invoicing.** Some industries, like construction, require customers to pay in installments. The process of sending several invoices for one project over the period of time is progressive invoicing. For farmers, I recommend you **select no.**

12. **Managing bills you owe.** One of the great features of QB is that you can manage your cash flow by prioritizing your bills. You may get several bills at a time, but not all are due on the same date. By entering them in QB, you can easily track which bills are due when, and use this tool to better manage cash flow. With great emphasis, I say, **select YES!**

13. **Tracking inventory in QuickBooks.** This question refers to inventory that you purchase (or grow) and then sell. To use this feature properly, you would need to enter in every case of tomatoes as you harvest them and then record when you sell them. It provides a tool for keeping track of how much product you have on hand. It does not include inventory of items you don't sell, such as seed or packaging (though you may still want to track them, just not in QB). QB inventory tracking is more complicated than it's worth. For this reason, I'd say **select no.**

14. **Tracking time in QuickBooks.** If you're new to QB, the learning curve may be steep and you're not ready to *track time*. As you get more advanced, you'll find this function useful. You can track your employees' and your own hours. It will be especially helpful in understanding your labor costs as it relates to the different products you grow, raise and/or sell. **Select yes.**

15. **Do you have employees?** W-2 employees have taxes withheld (by you, the employer). 1099 contractors are laborers who you pay regularly, but they pay their own payroll taxes. If you have people working for you on the farm, **Select yes.** (Just be forewarned, Intuit will try to sell you on its payroll services.[1])

Using Accounts in QuickBooks

16. **Select the date to start tracking your finances.** Are you just starting your business now? Select today's date or the first day of the month.

    If you've been in business for a while, but just getting set up in Quick-Books, it can be tempting to start tracking from January 1, regardless of the date, so you can have a complete year. This means entering all your old receipts and deposits from the beginning of the year to the current date. For most entrepreneurs, the desire is great, but the reality is that QB becomes a tedious chore if you're playing catch-up; best to start at the beginning of this month.

17. **Review income and expense account.** This is the start of your **Chart of Accounts** (COA), the structure by which QB will organize your income and expenses. The *recommended accounts* are based on the industry you selected in Question 2. However, many are irrelevant for a small diversified farm, such as "Agricultural Program Payments." In the next section, I will provide suggestions for how to customize your COA. For now, *un-click* accounts that don't align with your business.

Click **Go to Setup.**

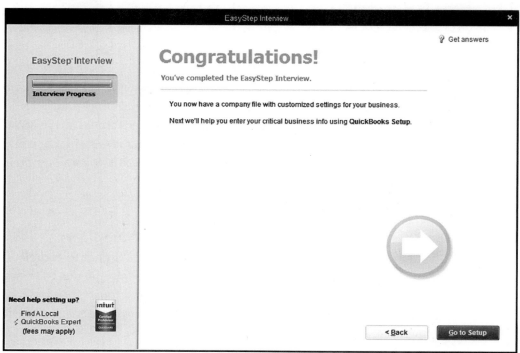

You've got a company file! Now add your info.

QuickBooks gives you the option to enter your customers, as well as the products and services you sell during the setup. My suggestion is that you skip adding customers now, and add them as you have transactions for them. Similarly, you can add *products and services* as you go (when creating invoices and tracking sales), and skip the initial setup here.

You will want to **Add your bank accounts** now, so you can enter expenses and make deposits. Most of your initial transactions will require a bank account.

   a. For the **Account Name,** type in the name you associate with the account. It may be "Farm Credit Checking" or "Business Checking." It doesn't have to match your bank statement.

   b. You do not need to enter your account number.

   c. The opening balance is the amount of money you have in your bank account today, assuming that all the checks have cleared. You can also enter the ending balance from your last bank statement. If you chose this option, you'll need to enter any transactions that happened between the closing bank statement and the day you are setting up your QB file.

   d. Chose the date of the opening balance. It could be today's date or the date of your last bank statement.

Create a separate line for all your business accounts.

In addition, I recommend setting up a "Petty Cash" bank account. This will give you the flexibility to track cash transactions as well as transactions that you accidentally make with your personal (not business) funds.[2]

The final window says, **"Now that you've added your account information, you can be even more productive using QuickBooks."** This is QB's first of many attempts to upsell you more products. Here, they want you to buy their checks. I suggest you decline.

And finally click **Start Working.**

QuickBooks Home Screen: An Overview of Things You'll Do

QB offers several options to access every task that you'll want to complete. You can access the menus from the top ribbon of pull-down menus (6) from the various icons around the home screen in the 5 sections (1–5) or from the sidebar shortcuts (7). And, of course, it offers many ways to upsell their products (8); I recommend you ignore those sections.

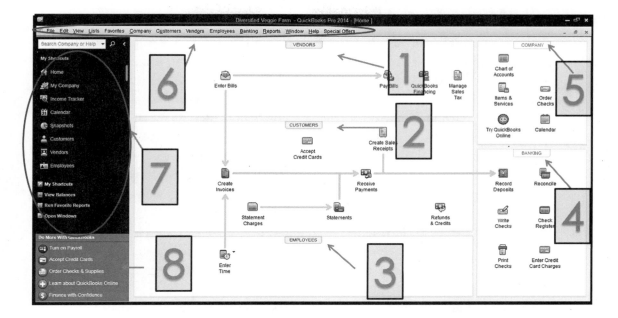

Section 1: Vendors

The vendors are people and companies to whom you owe money. Within this screen, there are three icons to pay attention to: **Enter Bills**, **Pay Bills** and **Manage Sales Tax**. Ignore QuickBooks Financing (they're just trying to sell you more products).

1. **Enter Bills:** When you receive a bill in the mail or from a vendor directly, such as the phone company, and you don't want to pay right away, you will log it in here. QB will now start tracking that bill and when it's due.

2. **Pay Bills:** When you pay a bill that was previously entered, you need to let QB know that you paid it. You do this by **paying bills**. You pay bills all the time, but only use this icon/function when you pay bills that you first entered (above in item 1).

3. **Manage Sales Tax:** If you've been collecting sales tax from your customers (not all farms do), you can click on this icon to manages the taxes you remit to the government.

Section 2: Customers

Customers are the people and companies who purchase from you. They could be your wholesale customers, CSA members, farmers market customers or folks who shop in your farm store. Not every customer is worth tracking separately. You may want to track all your CSA customers separately. But tracking the customers in your store or farmers market stall is laborious, and doesn't yield sufficient information to make it worth your while.

Within the customer section you can **create invoices** for customers who don't pay right away, record when you **receive payments** from customers for whom you wrote an invoice, **create sales receipts** for customers who pay straight away and **create statements** for customers with several outstanding invoices and assess them **statement (finance) charges**.

1. **Create Invoices:** For customers who don't pay you right away, you will create an invoice. This tells QB to start tracking who owes you money (which will show up in accounts receivable).

2. **Receive Payments:** If you create an invoice, then you need to let QB know it's been paid by using the **receive payment** icon. If you simply record the deposit (in the banking section) without "receiving" it first, QB will think the invoice is still unpaid.

3. **Statements:** If customers delay paying, or they accumulate several invoices in one month, you may want to create a **statement**. This will provide a summary of outstanding invoices. If they are late with payments, you may want to assess a few with **statement charges**.

4. **Sales Receipts:** There are (at least) two instances where you would use sales receipts: first, when a customer pays on the spot for their purchases and wants a receipt; second, when you want to record the details of your sales before depositing money into your bank account.

Section 3: Employees

1. **Enter Time:** This will allow you to enter and track the time you and your employees work.

Section 4: Banking

1. **Make Deposits:** When you go to the bank with checks and cash, or receive an automated payment from a credit card company (depositing funds resultant from customers paying with their credit cards), you will record them here. If you **received a payment** from an above-mentioned invoice in the customer section, you will record the deposit here. Note: In the dropdown menu (section 6), this function is labeled "record deposits."

2. **Reconcile:** Once a month, when you receive your bank statement, you

will reconcile your bank account to make sure you recorded all the transactions correctly. This will ensure that you have correctly tracked your bank balance, and can therefore trust QB to show the correct amount you have in the bank. You will also use the reconcile function when you receive your credit card statement.

3. **Write Checks:** Anytime you make a withdrawal from your bank account or record an expense paid by your debit card or check, you can enter it here. Even if you aren't actually writing a check, but recording a transaction from you debit card, enter it here.

4. **Check Register:** This looks more like a traditional old-fashioned check register. You can certainly enter transactions here (both deposits with withdrawals), but the interface is less convenient than the **make deposit** and **write checks** functions. The **check register** is convenient for scrolling quickly through past transactions to find something.

5. **Print Checks:** If your checks can be run through a printer, you can have QB set them up to print.

6. **Enter Credit Card Charges:** When you charge things to your credit card (as opposed to paying with your debit card, which withdraws directly from your checking account), you enter the expenses and purchases here.

Section 5: Company

This section is less about day-to-day activity and more about setup and account layout.

1. **Chart of Accounts:** This is the layout of your accounting structure. You can add or edit the accounting structure. Any time you make an edit to the accounting structure, it will flow through and update all past transactions.

2. **Items & Services:** Items and services are the things you actually sell: the tomatoes, chickens, flowers or landscaping services. They are a part of your revenue stream. Tomatoes are a part of wholesale or farmers market sales. Landscaping might be a part of a broader Maintenance Income category. In this section, you can set up and review your *items & services* and determine within which income or expense category they belong.

3. **Calendar:** You can see when you entered in transactions in a summary format.

4. **Order Checks** and **Try QuickBooks Online** are Intuit's attempt to upsell. I recommend you ignore these sections.

Sections 6 and 7: Top Ribbon Menu and Sidebar Shortcuts

All the options mentioned above, plus many more functions, can also be found through these series of pull-down menus.

Section 8: More Upselling

If you want to purchase more from Intuit, you can poke around in this section.

Setting up the Chart of Accounts (COA)

The COA is the structure you will use to organize all your revenue and expenses on the income statement, as well as the information for your balance sheet. The better organized, the better information you can extract.

During the initial setup, QB suggested sales and revenue categories. They are not meaningful categories for most small farms and function more for filing taxes than providing nuanced information about your business.

I recommend that you set up the COA in a way that makes sense for you and your business. You can give the accounts names that make sense to you and create broad categories to further organize your financial statements.

For each account you set up, you can have two levels of sub-accounts. For example, you can have a main account for CSA Revenue with sub-accounts for summer CSA, winter CSA and flower CSA. The sub-accounts allow you to organize information into more meaningful categories.

For the expenses, I suggest organizing them into five main categories:
- Direct Operating
- Labor
- General and Administrative
- Repairs and Maintenance
- Occupancy

All expenses should be a subcategory of one of these broader categories.

You can import a chart of accounts using the **QB Chart of Accounts** template (juliashanks.com/TheFarmersOfficeTemplates/) or you manually set them up. Instructions for importing the QB Chart of Accounts are included with the template. For a list of suggested accounts, see Appendix 3.

Customizing the COA

If you do not import a customized COA, you can customize from the home screen. Click on the **Chart of Accounts** icon.

Click on the **Chart of Accounts** icon to get to the COA set-up screen.

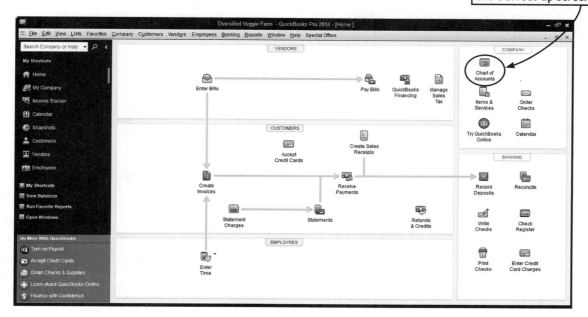

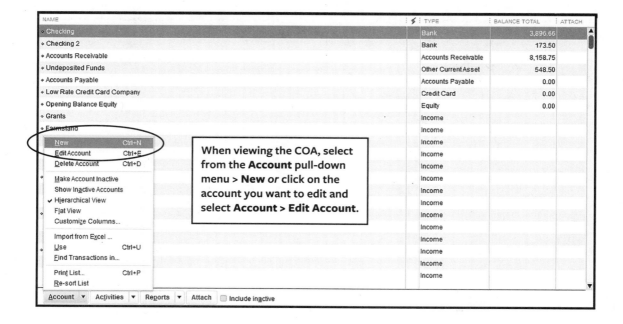

When viewing the COA, select from the **Account** pull-down menu > **New** *or* click on the account you want to edit and select **Account > Edit Account**.

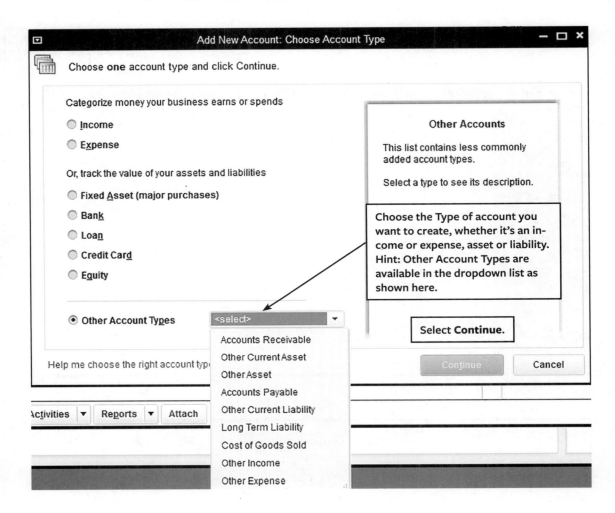

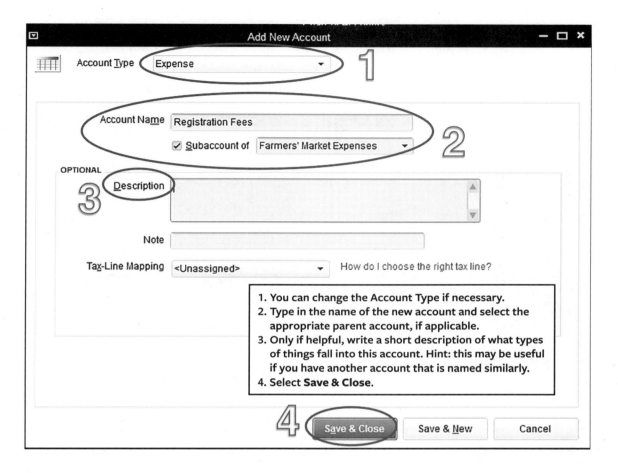

Editing Your Chart of Accounts

After entering a few transactions in QB, you may realize that your initial ideas on the layout of Chart of Accounts do not make sense. Maybe you want to rename accounts, or create different sub-accounts. Fear not; you can easily edit the accounts without changing individual transactions. Whether you rename an account or make it a sub-account of a different account, all previous transactions associated with the old account name will be modified to reflect the new settings.

Setting up Accounts and Vendors as You Go Along

With the basic setup of QuickBooks complete, you can begin entering transactions. The first few transactions may be bumpy. Besides the usual bumps of learning a new system, you will be adding vendors and customers as you go along. Each time you receive and enter a new bill, QB will not recognize new names and will ask if you want to set up an account.

It will look like this:

If you select **Quick Add** the name will be added to your vendor list. To add all the details about the vendor, including their full name and address, click on **Set Up** instead.

After you enter vendors and customers once, QB will remember them and autofill as you start typing.

Go to Chapter 7, Day-to-Day: Using QuickBooks for Cash Management to read about how to enter different transactions into QB.

Setting up Advanced Functions

As your business expands and becomes more complex, you'll want more insight from your numbers. QuickBooks can grow with you and provide more ways to slice and dice the data. Some advance functions include:

- Classes
- Customer: Jobs
- Time Tracking

Classes

When tracking revenue, it can be difficult to decide if you want to track by product (such as flowers and produce) or by sales channel (such as farmers markets and wholesale). If you set up revenue accounts for both produce *and* farmers markets, then you have the inevitable conundrum when you go to record the daily sales at the farmers market. Where does it go: under produce or under farmers markets?

By using classes, you can do both. You can set up the revenue/income category by sales channel and your classes by product type. Every time you enter a transaction, you can select the appropriate account category (for revenue and expenses) *and* class. Tracking by class allows you to take a cross-section of your business. You can see a mini profit and loss statement for each class.

To create classes, go to the drop down menu:

Lists > Class List

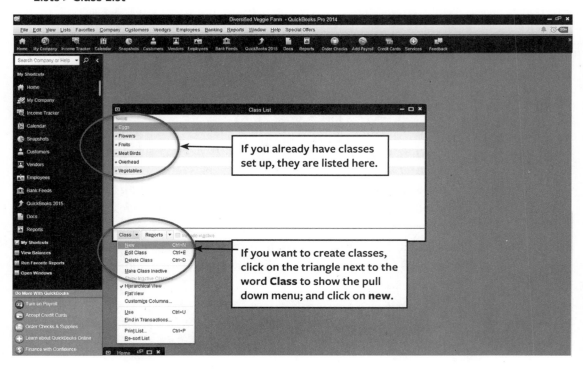

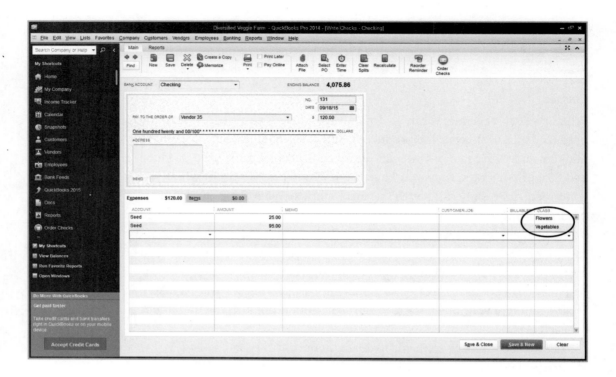

Let's say you buy seeds from a vendor; some are for flowers and some are for vegetables. By separating out the purchase into two lines, you can begin tracking seed expense, not just as a general category, but also what types of seeds—whether its for flowers or vegetables.

Why not just have sub-accounts under seeds for flowers and vegetables?

This is possible, however it will get messy. You may also have different packaging for veggies than for flowers. If you use sub-accounts, then you will have two more sub-accounts under packaging. And soil amendments: you may use something different for flowers than you do for vegetables. At the end of the year, when you want to total up the expenses for vegetables and for flowers, you will have to manually add up the numbers. If you use classes, then you can see both the total spent on seed as well as expenses in each class (see Table 6.1 on the following page for a sample Profit and Loss report by class).

TABLE 6.1. Sample Profit and Loss with Columns by Class Report

	Chickens	Eggs	Flowers	Pork	Vegetables	Overhead	Total
Ordinary Income/Expense							
Income							
CSA Shares	905.46	38,636.30	0.00	0.00	0.00	0.00	39,541.76
Farmers Market	599.00	6,643.50	3,983.00	1,538.25	14,632.50	0.00	27,396.25
On-farm Sales	1,106.00	118.00	60.00	859.79	0.00	0.00	2,143.79
Wholesale	0.00	12,406.50	0.00	0.00	0.00	0.00	12,406.50
Total Income	2,610.46	57,804.30	4,043.00	2,398.04	14,632.50	0.00	81,488.30
Expense							
Labor							
Volunteer Gifts and Meals	0.00	0.00	0.00	0.00	0.00	295.69	295.69
Work Gear	0.00	0.00	0.00	0.00	0.00	600.35	600.35
Contract Labor	343.85	0.00	0.00	0.00	0.00	2,286.74	2,630.59
Total Labor	343.85	0.00	0.00	0.00	0.00	3,182.78	3,526.63
Direct Operating							
Brooder	113.47	0.00	0.00	0.00	0.00	0.00	113.47
Coop	0.00	4,715.41	0.00	0.00	0.00	0.00	4,715.41
Feed	1,048.58	13,767.85	0.00	2,537.67	0.00	0.00	17,354.10
Livestock	361.50	3,750.00	0.00	800.00	0.00	0.00	4,911.50
Seeds and Plants	0.00	0.00	528.42	0.00	2,537.94	0.00	3,066.36
Supplies	97.06	70.00	33.98	188.39	836.18	856.76	2,082.37
Propagation	0.00	0.00	0.00	0.00	688.42	0.00	688.42
Fert and Ammendments	0.00	0.00	0.00	0.00	1,985.04	0.00	1,985.04
Packaging	0.00	987.17	74.28	0.00	0.00	0.00	1,061.45
Storage	0.00	0.00	0.00	0.00	1,091.78	0.00	1,091.78
Mulch	0.00	0.00	0.00	0.00	60.00	0.00	60.00
Small Tools and Equipment	0.00	0.00	0.00	0.00	24.00	255.97	279.97
Supplements	0.00	307.64	0.00	0.00	0.00	0.00	307.64
Winter Supplies	0.00	284.44	0.00	0.00	0.00	0.00	284.44
Direct Operating—Other	1,213.75	0.00	0.00	0.00	76.99	0.00	1,290.74
Total Direct Operating	2,834.36	23,882.51	636.68	3,526.06	7,300.35	1,112.73	39,292.69
General Administrative							
Bank Service Charges	0.00	0.00	0.00	0.00	0.00	33.40	33.40
Computer and Internet	0.00	0.00	0.00	0.00	0.00	181.99	181.99
Dues and Subscriptions	0.00	0.00	0.00	0.00	313.00	104.00	417.00
General Upkeep	0.00	0.00	0.00	0.00	0.00	80.07	80.07
Insurance	0.00	0.00	0.00	0.00	0.00	250.00	250.00
Interest	0.00	0.00	0.00	0.00	0.00	24.00	24.00
Licenses and Fees	0.00	0.00	0.00	0.00	75.00	1,036.50	1,111.50
Office Supplies	0.00	0.00	0.00	0.00	0.00	405.24	405.24
Professional Development	0.00	0.00	0.00	0.00	0.00	412.49	412.49
Rent	0.00	0.00	0.00	0.00	0.00	4,980.00	4,980.00
Taxes—Property	0.00	0.00	0.00	0.00	18.00	3,202.00	3,220.00
Utilities	0.00	0.00	0.00	0.00	0.00	1,390.87	1,390.87
Total General Administrative	0.00	0.00	0.00	0.00	406.00	12,100.56	12,506.56
Fuel							
Equipment	0.00	0.00	0.00	0.00	0.00	15.07	15.07
Vehicles	0.00	0.00	0.00	0.00	0.00	2,618.33	2,618.33
Total Fuel	0.00	0.00	0.00	0.00	0.00	2,633.40	2,633.40
Marketing	0.00	118.26	57.40	0.00	0.00	411.55	587.21
Car and Truck Expenses	0.00	0.00	0.00	0.00	0.00	815.92	815.92
Repairs and Maintenance	0.00	0.00	0.00	0.00	120.00	270.18	390.18
Total Expense	3,178.21	24,000.77	694.08	3,526.06	7,826.35	20,527.12	59,752.59
Net Ordinary Income	−567.75	33,803.53	3,348.92	−1,128.02	6,806.15	−20,527.12	21,735.71

Customer: Jobs

If you sell to wholesale clients, you'll want to know how much they buy and how profitable they are as clients. **Customer: Jobs** allows you to track your revenue by each customer. In addition, you can use **Customer: Jobs** to track farmers markets sales, not individual customers but the market as a whole by creating *customers* for each market you attend.

In addition to tracking revenue, you can mark expenses that are specifically related to a customer or market. While many expenses are more general—like truck maintenance or animal feed—some may be specific. If you custom grow for a client, then the seeds expense can be linked to that customer. If you pay vendor fees at a farmers market, that expense can be linked to that market. If you fuel up your truck to deliver wholesale, you can track this as both a fuel expense and a wholesale expense.

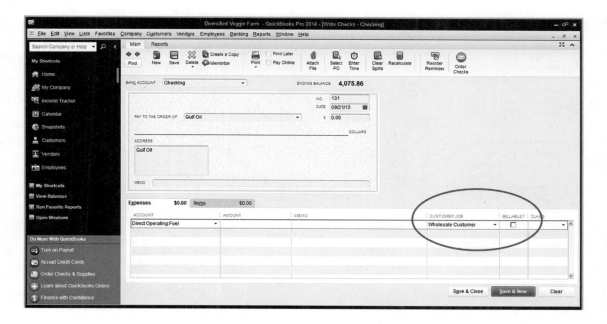

For every transaction, there is a column to enter a customer. Be sure to uncheck the box in the billable column. If you leave it checked, QB will add it to future invoices; unless your customers reimburse you for expenses, you don't want to do this.

Time Tracking

The time tracking function allows you and your employees to enter hours worked; you can simply use it as a time card.

As you get more advanced, you can also track time spent working on different crops and tasks. For example, you may want to track how much time you spend harvesting tomatoes or weeding. Because QB only allows the option to track time by customer (and not class), using **Customer: Jobs** for time tracking is a work-around. **Customer** could be a specific crop (such as tomatoes or kale). When you harvest tomatoes, you can enter the time with the **Customer: Job** (tomatoes) and the **item** as "harvesting."

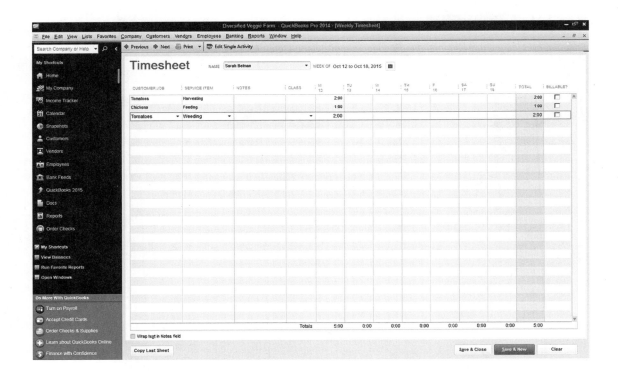

Items

Within each account category (whether it's an expense or revenue), you may want to track more details. For example, on an invoice for a wholesale customer, you want to list that you sold five pounds of bok choy, one case of chard and six bunches of radishes (as opposed to the more generic **whole-sale vegetables** which is how you list sales on the profit and loss statement). These **items** are part of wholesale revenue. This also provides a way to track another cross-section of your business data. It's helpful if you want to see how many pounds (or cases) of a particular crop you sold in a season. You can access this information in Reports > Sales > Sales by Item Summary.

When you set up **items**, you will need to assign a revenue category. For example, you may have bok choy, chard and radish as **items**, and they will all be a part of the **wholesale** revenue account. To set up **Items**:

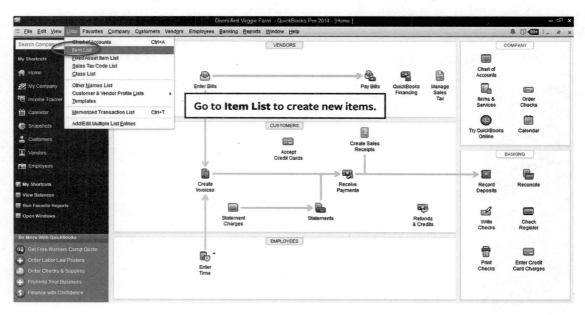

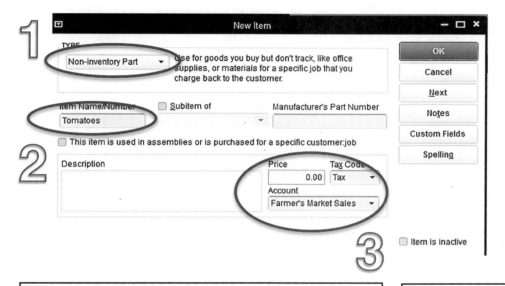

1. For **Type** select **Non-Inventory Part**.

2. **Item Name/Number** is what will appear on invoices and sales receipts. Choose a name that is short and descriptive.

3. If your **Price** is consistent, you can enter it here. If not, you can leave it at zero and adjust the price on invoices and sales receipts.

   Select the revenue account associated with the item.

Note: If you sell an item in multiple channels—such as selling tomatoes to both wholesale and farmers market customers, you will need to create two different items. To distinguish between the two, you can add a prefix to the **Item Name/Number** such as "W" for Wholesale and "FM" for Farmers Markets

Reports in QuickBooks

QuickBooks can create dozens of reports—each providing a different snapshot about your business. Each report can be further customized to make sure you're getting the information you need. Noodle around the reports section to see some of the options. Here are six reports that you'll want to become familiar with and where to find them in the submenus of Reports.

Reports > Company & Financial > Profit and Loss Standard

This is your basic income statement, and you can track the profitability of your business. For a deeper look, you can *customize* (on the Mac you will click on Options) the date range, compare the profit and loss to previous years and look at subsets of your business (by class or by job).

Reports > Company & Financial > Balance Sheet Standard

The balance sheet gives you an overview of your assets and liabilities. This can be especially helpful to see in a single snapshot the cash you have on hand plus the cash you expect to come in from customers (A/R). You can use this as part of your regular cash planning to look at your accounts payable, credit card debt and upcoming expenses.

Reports > Vendors and Payables > A/P Aging Summary

As your business grows, so does the volume of bills you receive. It can be overwhelming to track who you owe and when. The A/P Aging Summary provides a quick snapshot of the bills you owe, and which, if any, are overdue and by how long.

Reports > Customers and Receivables >A/R Aging Summary

With wholesale clients, you may choose to extend terms to them, meaning they do not pay you right away. You can get a quick snapshot of who owes you, and if they are late paying, through this report. This is especially important when you're tightly managing cash flow. First, you can see what money is coming when. Second, if you have customers who are falling behind on their bills, you know who to call.

Reports > Jobs, Time Mileage > Time by Job Summary

If you track your time by job or by crop, you can see an overview of how much time you spend on different tasks.

Reports > Jobs, Time Mileage > Time by Name

If you track your employees hours in QB, you can see a report of how many hours they work within a given time period.

QuickBooks Terminology

Many of the terms used in QB are familiar words from everyday life, but the way QB defines them may be different from how you do. We've discussed most of these already, but here's a refresher of these terms.

Accounts: These are the categories of revenue, expenses, assets and liabilities. These are *not* bank accounts. Each subcategory (farmers market sales, soil expense, fertilizer) is an account. QB uses the accounts to organize all your transactions into your financial statements.

Sales Receipts: These are sales receipts that you give your customers (and *not* receipts you receive from your vendors). If a customer pays cash and wants a receipt for their records, you can create a sales receipt to record the revenue and provide the paper trail for your customer.

Invoices: These are statements you provide to your customers who do not pay right away. They are not the bills that you receive from vendors. While they are still "invoices" for the purposes of QB, what you receive from your vendors are **Bills**.

Classes: These are subcategories of your business and a way to segment different lines of business. If you track revenue by sales channel (farmers markets, CSA, wholesales, etc.), classes allow you to track revenue and expenses by product. If you have multiple segments of your business, such as landscaping and product sales, classes let you separate revenue and expenses by these two segments. Most reports can be created with a breakdown by class.

Debits and Credits: Intuitively, we think of debits as decreases in cash and credits as increases: our debit card decreases cash in our bank account, and a credit gives us money back. In QuickBooks, as in proper accounting, they simply refer to the left and right columns in **general journal entries** (and double-entry accounting). We'll discuss **journal entries** in the next chapter.

Notes
1. Intuit is the software company that created and sells QuickBooks.
2. You do keep your personal finances separate from your business, right??

Day-to-Day: Using QuickBooks for Cash Management

Do you remember Aesop's fable of the Grasshopper and the Ant?

In a field one summer's day a Grasshopper was hopping about, chirping and singing to its heart's content. An Ant passed by, bearing along with great toil an ear of corn he was taking to the nest.

"Why not come and chat with me," said the Grasshopper, "instead of toiling and moiling in that way?"

"I am helping to lay up food for the winter," said the Ant, "and recommend you to do the same."

"Why bother about winter?" said the Grasshopper, "We have got plenty of food at present." But the Ant went on its way and continued its toil.

When the winter came the Grasshopper found itself dying of hunger, while it saw the ants distributing, every day, corn and grain from the stores they had collected in the summer.

Then the Grasshopper knew: It is best to prepare for the days of necessity.

When it comes to food, you probably think of yourself as the Ant: In the summer, you can tomatoes, freeze berries and meat, and pickle vegetables. Just as you preserve the summer's bounty to tide you through the winter, you must also save the cash you earn during peak season for the inevitable slow periods.

While most businesses have some degree of seasonality, few have such wide swings as a farm—from the depths of winter when the fields are frozen to the peak of summer when tomatoes and eggplant fill every harvest bin and cover every inch of your packing room.

In order to have enough cash to pay bills in the winter, and avoid unnecessarily running up your credit card or taking out a loan, you need to set aside money. It's too tempting to look at your bank balance in the summer and see a large number. I'm sure you could spend the balance five times over.

Staying on top of the day-to-day

Rachel called me in the early spring to help with her finances. She had been in business for about 10 years, but just couldn't get ahead on her credit card debt. In summer, business was slamming-jamming. As her bank balance grew, she aggressively paid down her credit card; she knew those high interest rates would be more burden than her business could bear. But by winter, she was short on cash, and charging up her credit cards again. She just couldn't figure out how to get out of this cycle.

In order to get a handle on her cash flow cycle, so she could avoid credit card debt, we needed to better understand the ebbs and flow of her cash—when the cash comes into the business and when it leaves—so that we could create a plan. And to understand the cycle of her business, she needed to get systems in place, specifically, setting her up in QuickBooks. The QuickBooks would help with the day-to-day management of cash. After Rachel had a few months of record-keeping under her belt, we'd start to see trends of the ebbs and flow of cash, the profitability of her different product lines, and then we could create a budget. The budget would be the guiding star of how she would make the numbers work: when could she make big purchases, and how much of her debt could she afford to pay down every month.

She had learned the hard way that, with the seasonal nature of her business, she couldn't just pay down her bills when she had cash. When she did, there was nothing left for the off-season.

There's the new tractor implement you've been eyeing, or the line of credit you want to pay down. But when winter rolls around, and you need to pay the rent, you want to make sure you have enough squirreled away.

Managing cash flow isn't difficult; but it's a process that takes commitment and consistency. Instead of always struggling to keep up with your bills, effective management allows you to get ahead and plan for growth. A daily routine can help you avoid a cash crisis, and navigate out of one if you accidentally find yourself there.

Managing cash flow is one part daily tracking and one part long-term planning. When you create a yearly budget that works, you need to monitor your cash on a regular basis to ensure you stay on track. You need good historical records (that come from daily tracking) to create a solid long-term plan.

To get the full benefit of long-term planning, start with an annual budget. Through the course of the year, and on a regular basis, you can use your budget to make sure you're on track, know when you're getting ahead and have the tools to make on-the-spot decisions that creep up (like that used-tractor-that-you-just-saw-and-is-totally-in-your-price-range-and-you-want-to-buy-it-now), but will it fit in your budget?

Creating a budget is much the same as creating monthly profit and loss projections (see chapter 4, The Business Planning Process). For internal planning purposes, I recommend a mash-up of an income statement and a cash flow statement. List all the revenue you think you earn each month, along with all the expenses and cash outflows you expect, including capital purchases. See Quick and Dirty Cash Flow Template at: juliashanks.com/TheFarmersOfficeTemplates/

Every Day—Data Entry (10 minutes)

At the end of the day, when you're pouring a drink and winding down, and before you jump in the shower, empty your wallet and spend a few minutes entering your expenses and income. The goal is to make data entry a part of your regular routine—just as you feed the chickens or irrigate your crops. With regular maintenance, you can better troubleshoot solutions when troubles arise. And at the end of the year, when you're filing taxes and planning ahead, the process will be exponentially easier.

1. Receipts in Your Wallet

Did you go to the store to pick up supplies or fuel up the truck? It's important to track your expenses before the receipts pile up (and it becomes overwhelming) and before you forget what you spent the money on.

    a. Empty your wallet of receipts, and divide them into three piles: one pile is for things you put on your charge card, the second pile is for things you put on your debit card, and the third pile is for things you paid in cash.

    b. Go to **Enter Credit Card Charges:** Enter in all the charges you put on your credit card (not your debit card). Be sure to track the right account and classes.

Enter Credit Card Charges

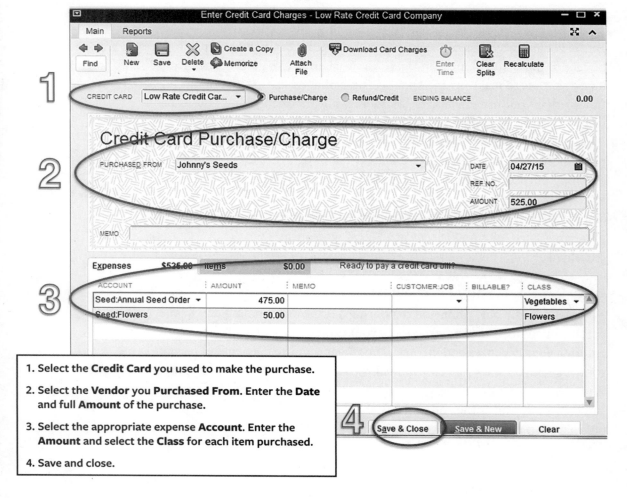

1. Select the **Credit Card** you used to make the purchase.

2. Select the **Vendor** you **Purchased From.** Enter the **Date** and full **Amount** of the purchase.

3. Select the appropriate expense **Account.** Enter the **Amount** and select the **Class** for each item purchased.

4. Save and close.

c. Go to **Write Checks:** Even though you are using the **write checks** function, think of this as your standard old checkbook register. You will enter debit card receipts and cash receipts with the same function. For cash transactions, be sure to select the **Petty Cash** account in Step #1.

Write Checks

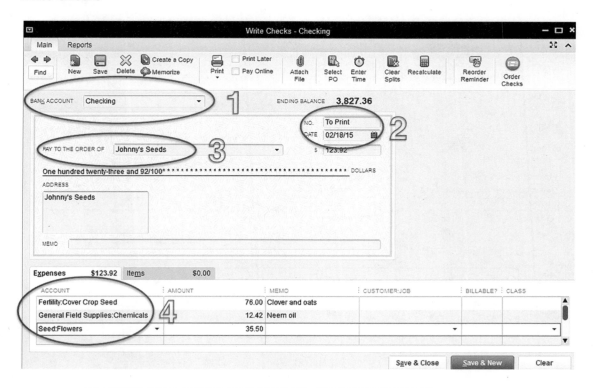

1. Select the **Account** from which you will be paying.

2. Select today's **Date** and the corresponding check number (for entering historical data, choose the **Date** on which the bill was paid and don't worry about entering the check number).

3. Select the **Vendor** that you are paying.

4. In the **Expenses** tab, select the appropriate expense **Account** that each purchased item falls into.

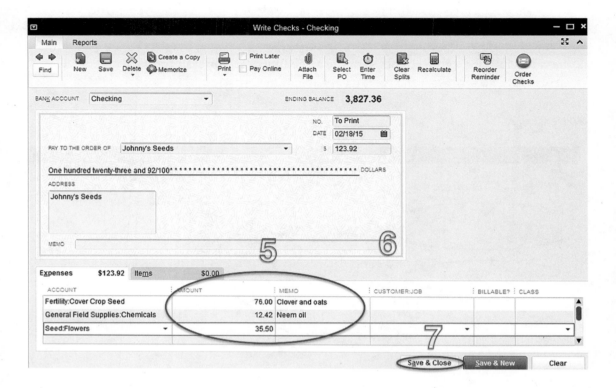

5. Type in the payment **Amount** and, if applicable, enter a brief **Memo**.

6. As applicable, assign the appropriate **Customer:Job** and **Class**.

3. **Save and Close**.

2. Enter and/or Pay Bills

Did you get the phone bill or cable bill today? Log the bills using the **Enter Bills** function. Track the day the bill was received as well as the day it is due. This will help you manage cash flow.

Enter Bills

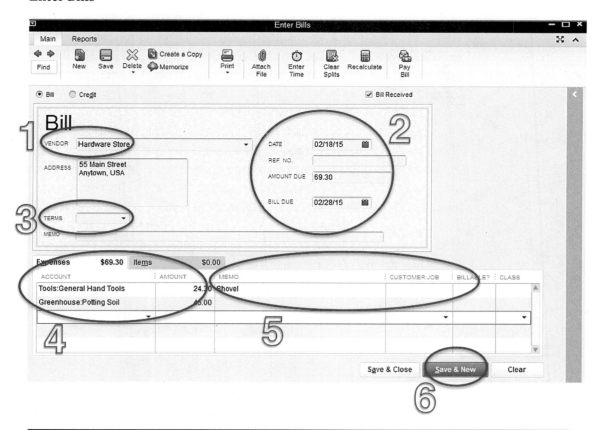

1. Enter the name of the **Vendor**. Once you've entered a vendor once, the information should auto-fill.

2. Be sure to enter the **Date** of the invoice as well as the **Bill Due** date.

3. Select the **Terms** of the bill, as applicable.

4. Under the Expenses tab, Select the proper **Account**. If you are unsure, use the drop-down menu. If purchases fall into two or more categories use a separate line for each category and allocate the dollar amount on each line.

5. Only enter **Memo** and **Customer Job** if applicable. Don't worry about the Billable column.

6. Click **Save & New** if you have more bills to enter. Otherwise, **Save & Close**.

3. Create Invoices, Enter Revenues and Record Receipts

Were you at the farmers market today? Did you get a check in the mail from a wholesale customer? Did you make wholesale deliveries? Did you deposit money into your bank account? The third task is to record the revenue you earned and the money that came in.

a. Create Invoices

Some customers may not pay right away when they receive your products. They will want an invoice (for their bookkeeping records); and you'll also want to create one to track who owes you money. The invoice details how much the customer owes you, by when (the terms) and for what.

Ideally, you create the invoice in QB before making your delivery run and hand it to the customer along with the product. Logistically, many farmers don't like writing invoices before a delivery; after all, the customer might add on a case of arugula to their order last minute. If you use QB online, you can create it on the spot (while making the delivery) and have it emailed. If you use the desktop version, you can leave your customer with a handwritten invoice when you deliver, and then create the invoice in QB when you return to the farm. Even if you do not send the QB invoice to the customer, you should still create a rudimentary one for your record keeping purposes.

Create Invoices

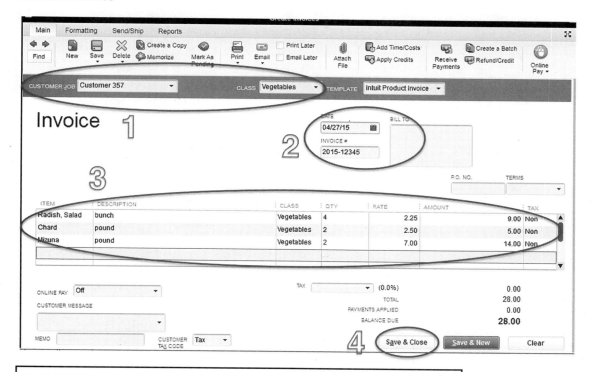

1. Select the appropriate **Customer** (and **Job** if applicable). Select which **Class** the customer fits (skip if you don't yet have **Classes** set up).

2. Select today's **Date**. If you previously sent invoices to this customer, **Customer** and **Bill To** address will auto-populate. Otherwise, fill these in manually.

3. Input the **Items** sold and **Quantity** of each. Add the price per unit under **Rate**. **Total Amount** will auto-calculate.

4. **Save & Close.**

b. Record Payments Received for Invoices

Did you receive checks in the mail to pay off invoices that you previously sent customers? Before you record the deposit, let QB know the customers paid by using the **Receive Payments** function.[1]

To receive payments, click on the **Receive Payments** icon on the home screen. This lets QB know that the customer no longer owes you money. The physical check may go into your wallet, QB will note the value of the payment in an account named **Undeposited Funds**.

Record Payments Received

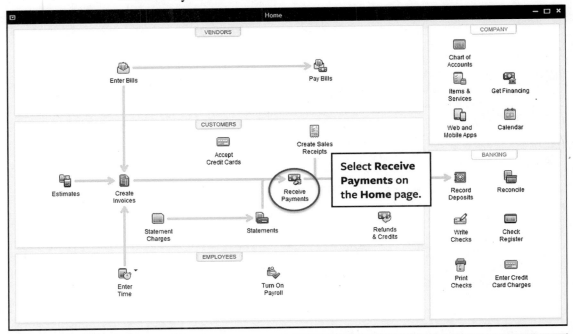

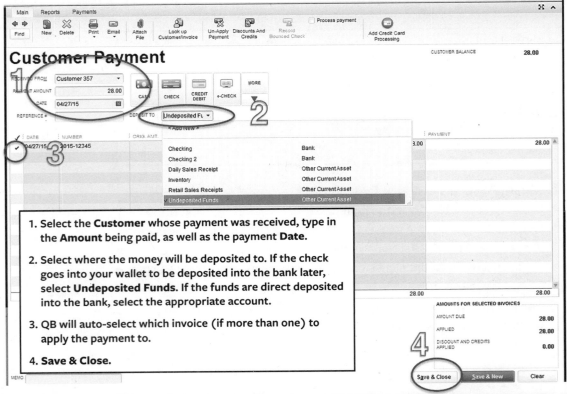

1. Select the **Customer** whose payment was received, type in the **Amount** being paid, as well as the payment **Date**.

2. Select where the money will be deposited to. If the check goes into your wallet to be deposited into the bank later, select **Undeposited Funds**. If the funds are direct deposited into the bank, select the appropriate account.

3. QB will auto-select which invoice (if more than one) to apply the payment to.

4. **Save & Close.**

Easy.

Note: QB wants to know when the check leaves your wallet and enters your bank account. When you record deposits in QB, you will have the option to select checks that have been *undeposited*. In the first window, you can select the checks that you are taking with you to the bank. (See details and screenshots below for more details on depositing funds.)

c. Record Daily Sales from CSA Customers, Farmers Market and Farm Store
Beyond receiving checks from wholesale customers (and others whom you invoiced), money comes into your business each day (farmers market sales, farm stand sales, checks received in the mail). It comes in all sorts of payment methods (checks, cash and credit cards) for all sorts of products (eggs, dairy, produce, flowers, etc.). The money enters your bank account in different categories than which you received it.

At the farmers market, you receive cash and credit cards. The credit card company will deposit the money into your bank account directly (less their fees). This deposit includes different classes of products (vegetables, eggs, flowers and/or meat birds). It may also include credit card sales from the farm stand, wholesale customers or CSA members. You may use some of the cash from the farmers market to make purchases on your way home. The remaining cash in your wallet could be deposited into the bank along with cash from the farm stand.

With this mash-up of cash flow and sales, it can be difficult to sort how you actually earn revenue.

In order to keep track of how you earned money (for management purposes) *and* how it's deposited into the bank (so you can reconcile your bank account and manage cash flow), you need to track incoming funds in a two-step process. In the first step, you record the sales. In the second step, you record the deposits. This allows you to fully categorize income by revenue stream (farmers markets, CSA, farm stand, etc.) as well as by product type (or class, such as produce, flowers and meat).

If you opt against this two-step method, you won't be able to track the different types of revenue beyond the generic **Farm Sales**. This can be especially problematic for reporting to GAP (Good Agricultural Practices) and the organic certification boards.

Side Note: When you go the bank, you may have some of this mash-up money, as well as checks that were *received* from Wholesale Customers (which you recorded in Step 2). When you click on **Make Deposit** and open up the deposit window, QB will ask if you want to deposit other checks received. The checks listed in this window will be the checks you noted were received to pay off invoices and/or sales receipts.

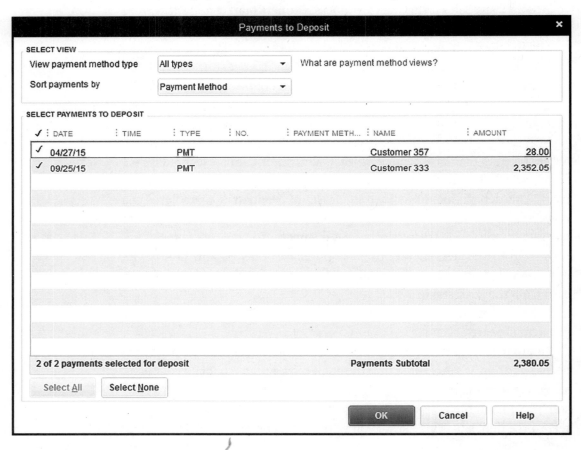

This is the process in QuickBooks:

Step 1. Note the sales by revenue stream and by class by creating a **journal entry.** A journal entry only allows you to track by class. If you want to track items as well, then you can enter sales using a **sales receipt.**

1. Receive money by sales channel

CSA

Farm stand

Wholesale

Received Money
(Daily Sales Receipts/
Undeposited Funds)

Credit Card Deposits
Checks
Cash

**2. Deposit money
as it goes into
the bank**

When a customer pays you, whether by credit card, cash or check, the money doesn't automatically appear in your bank account. It's delayed: either it sits in your wallet until you physically take the money to the bank, or the credit card company deposits it. Either way, you earned the money, it's yours. It's just not in the bank.

At this point, it's easy to sort how the money was earned. You can track whether it came from the farm store, CSA or farmers market; as well as tracking whether it was vegetables, meat, dairy or eggs.

In QuickBooks, you want to record the money received (though not deposited) in an asset account,[2] either **undeposited funds**, which is one of the QB automated accounts, or you can **deposit** it into a temporary holding account called **Daily Sales Receipts**.[3] Daily Sales Receipts and undeposited funds are just another way of saying you have the money in your wallet or cash box, but haven't yet deposited it into your physical bank account. (For further tips on how to track the sales of specific classes of products from the farmers market or farm store, see Inventory Tracking in Chapter 8.)

To create a journal entry:

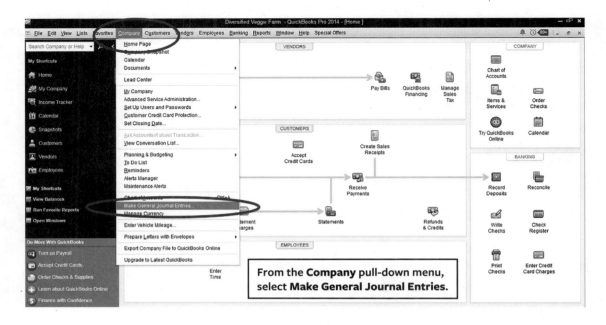

From the **Company** pull-down menu, select **Make General Journal Entries.**

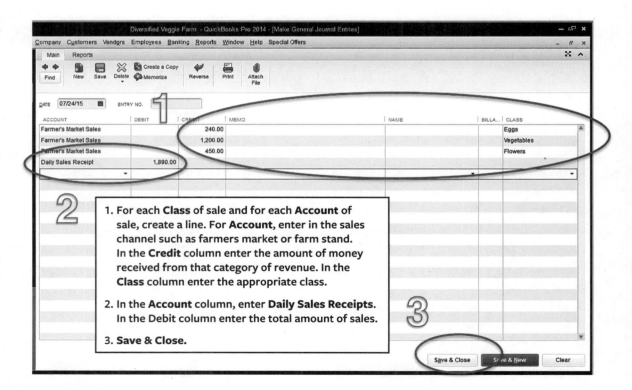

1. For each **Class** of sale and for each **Account** of sale, create a line. For **Account**, enter in the sales channel such as farmers market or farm stand. In the **Credit** column enter the amount of money received from that category of revenue. In the **Class** column enter the appropriate class.

2. In the **Account** column, enter **Daily Sales Receipts**. In the Debit column enter the total amount of sales.

3. **Save & Close.**

Step 2. Record the deposit into your bank account.

When the money is deposited into the bank, because you physically took it there or it was automatically deposited from the credit card company, you can record the deposit in QB. The money is *transferred* from the **daily sales receipts** account into your bank account. The money goes from being undeposited to deposited. Cash that doesn't make it into your bank account is *transferred* into your petty cash account. This cash will likely be used to pay future expenses.

Because you already recorded the sales categories in the previous step, this is just marking the deposit.

When the money gets deposited into the bank, record the deposit. In the **Account** column, note that the money is from **Daily Sales Receipts**.

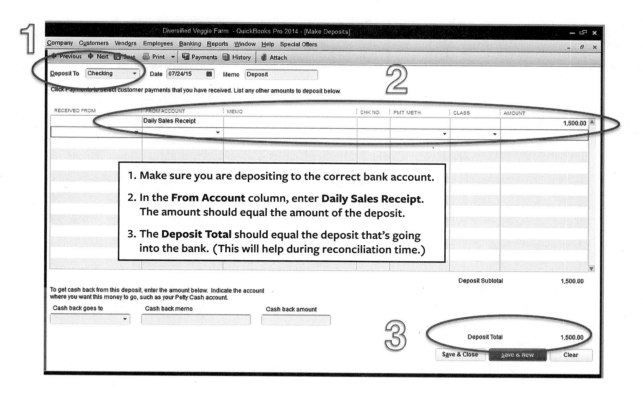

Be sure to select the checks that you are depositing. This lets QB know that the money is no longer in your wallet and is in the bank (it will also prevent you from double counting sales).

Further, if you used a **sales receipt** to record farmers market or other sales, then list the funds in this window as well. It may be easier to deposit the farmers market sales into your **petty cash** account, and then transfer funds into your bank account when it happens. As you know, not all the money you receive from the farmers market makes it into your checking account.

The deposit total in QB should match the actual deposit amount—this will make reconciling your bank account much easier.

This can be confusing. If you need to ponder this for a bit, that's okay. Go do some weeding, take a walk or have a drink. Once you get the hang of it, these daily QB chores will take less than 10 minutes. I promise.

The day-to-day of QB can seem overwhelming; especially at first when everything is new and not intuitive. Rachel felt the same way; and as the hectic spring headed into summer, she just could not commit the time to this regular task. A few weeks after the initial setup, she called me and spent 20 minutes explaining why it just wouldn't work now. It was her busy time, and she needed to focus on making as much money as she could.

Based on her pattern of running low on cash in the winter, I knew something was amiss. She could not afford to wait until the fall, her slow season, to start her cash planning; by then she'd be headed into the same cycle of borrowing cash. I knew that if she could commit to QB for two to three months, she would start to see the results; she would clearly see what was making her money and what was not.

Without a budget during the height of the season, it's easy to spend, spend, spend. And when winter comes, they'll be no money squirreled away.

Every Week—Pay Bills and Evaluate Cash Balance (30 minutes)

Whether or not you use QuickBooks, you must pay your bills. With QB, this regular task can be used as a time to evaluate your cash balance, which can be particularly important during periods with low sales.

I recommend that you designate a regular time for paying bills and cash management, a time when you can work uninterrupted. Perhaps it's Tuesdays at 3 PM, or maybe Fridays at 6 AM before the crew shows up. Find a time when you can regularly commit to 30 minutes. Some farmers prefer to do this once every two weeks. In theory, that's okay; though it's harder to make it a habit.

The goals are to pay the bills due in the coming week and take a gander at your cash balance. The more regularly you review your cash balance, the quicker and more agilely you can make adjustments. Do you have the cash to pay all your due bills? Can you afford to pay off credit cards or purchase new equipment? How much can you save to purchase the disc harrow you've been eyeing?

1. Pay Bills Function

Go to the **Pay Bills** function and select the date range of due on or before one week from today's date.

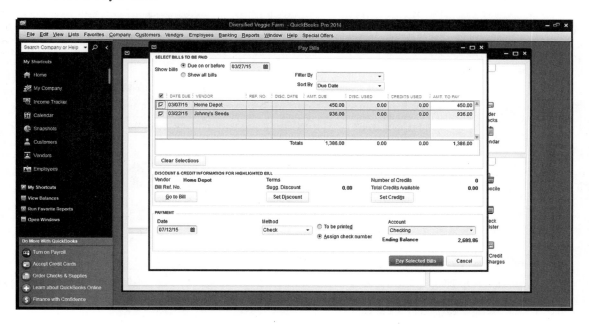

Managerial Thinking

   a. If you opt to pay all the bills at once, you'll see the total amount of bills being paid. In this example, the total is $1,386.

   b. As you check off bills to be paid, the ending balance will reflect what you'll have in the bank after the checks are written and the bills are paid. The bank balance shows $2,689.86 if all bills are paid.

c. Decision: Is $2,689.86 enough of a cushion in the bank?
   • What cash do you expect to receive in the next week?
   • What other bills do you expect to owe in the next week?

2. Outline Cash Inflows and Outflows for Coming Week

In order to make the above decision, you need to gather information.

a. What other payments do you have this coming week?

These are bills that you didn't enter into QuickBooks but you know are coming up. It could be a mortgage payment, equipment you're about to take in for repair, payroll, or the weekly farmers market fee. Maybe you plan to purchase new boars and sows.

b. What money do you have coming in?

Are you attending a farmers market? How much cash do you expect to earn? Are you expecting payment from customers? Check your **Accounts Receivable [A/R] Aging Summary** in **Reports**.

c. How much money do you set aside each week for slow periods, annual taxes and unexpected expenses?

If you haven't created a cash flow budget, you can start with $250 per week. I recommend that you set up a separate savings account with automatic transfers. If you can't see the money in your checking account, it will be easier to save it.

In the example in Table 7.1, extra cash is available. This can be used to pay down debt, if you have any, or set aside for future growth or unexpected emergencies.

TABLE 7.1. Example of Weekly Cash Analysis—with Extra Cash

Beginning Bank Balance	$4,075
Bills Due This Week	–$1,386
Other Payments This Week	–$750
Expected Cash Inflow	$2,500
Cash to Set Aside for Taxes and Slow Periods	–$250
Available Cash	$4,189

At the end of the month, there are often more expenses than in the middle: the mortgage (or rent) payment is due the same week as payroll, and the Visa bill is expected too. In the example in Table 7.2, there isn't enough cash available to cover all the cash needs. As you consider your options, look

TABLE 7.2. Example of Weekly Cash Analysis—with Deficit

Beginning Bank Balance	$4,075	
Bills Due This Week		−$1,386
Other Payments This Week		−$5,750
Expected Cash Inflow	$2,500	
Cash to Set Aside for Taxes and Slow Periods		−$250
Available Cash		−$811

back at the cash flow budget you created. Did you project revenues would be low and expense would be high? If this is an example cash analysis for April, then the numbers may make sense. Otherwise, look more closely:

- Are your expenses in line with your budget, or did you overspend?
- Are your revenues as high as you anticipated? If sales are off, why? What can you do to bring them back up?

If revenues and expenses are in line with your projections, and this is a slow period for which you planned, you can dip into the reserves you set aside for such a time. If not, then you need to evaluate the changes you can make in the following weeks.

If your projected cash balance is insufficient to cover the weekly bills, then first and foremost get through the next week without bouncing checks or accruing interest and finance charges. These kinds of fees can add up and start a downward spiral. I've seen farmers regularly pay $50 to $100 a month in fees, or $600 to $1,200 in one year. I'm sure you can think of many other ways you'd rather spend your money!

To get through the next week or two, you may need to delay payments for a few days on some bills. In order to stay on track long term, you need to compensate in the following weeks. What can you do to trim expenses? What can you do to increase revenues?

Let's be clear: One week exceeding the projected expenses in your budget is not a deal-breaker, but if it happens regularly, then a) you need to reassess your budget, and b) you will be in danger of getting into a hole that will be difficult to climb out of.

One comment I hear most often in these types of situations is "I had to spend the extra $500. If I didn't then I wouldn't have the supplies I needed to start the season." Go back and read the previous paragraph.

That said, sometimes tough choices need to be made.

Dealing with the Unexpected

The winter of 2014–2015 smashed snowfall records in New England with a seasonal total of more than 100 inches and close to 65 inches in February alone. The persistent cold and snow collapsed greenhouses across the region. When the snow finally melted in April, Farmer Steve had little time to rebuild the greenhouses before he needed to start seedlings and plantings; if he waited a moment longer, he'd fall behind for the rest of the growing season, both in terms of production and the almighty cash flow. Steve is a seasoned farmer, with 15 years' experience under his belt. He knows that spring is traditionally a tight cash period, and this unexpected expense was going to set him back even further. He had to plan carefully if he wanted to avoid getting into a deep hole. In addition to holding back on unnecessary purchases, he ramped up the marketing for his CSA. This expedited the early spring cash infusion. He was also able to more aggressively push his storage crops that he had been saving for the early spring farmers markets to wholesale customers.

Things happen. Even the most carefully planned ventures can have unexpected turns. Most of the time, it's not as dire as the spring of 2015. This is just one more reason why it's so important to create a budget, stick to it and have a backup plan for the unexpected.

Calculating Your Monthly/Weekly Savings Target

How much money do you need to set aside each week or month? In the first year of business, it may be hard to gauge. How slow will the slow periods be? How high will your taxes be? In subsequent years, you'll have a better clue. Either way, you should create a monthly cash flow budget (see Quick and Dirty Cash Flow Template) and create a plan for setting aside cash.

Process

1. Fill Out the Quick and Dirty Cash Flow Spreadsheet. Make sure that you include:

    a. a salary or draw for yourself so that you can pay for your living expenses

    b. monthly or quarterly tax payments

    c. planned purchases and debt repayment

2. Evaluate the year—does the year end with a positive cash balance? If yes, then proceed to the next step. If not, reevaluate your budget. Where can you increase revenues and decrease expenses?

3. Is there a month where you end with a negative cash balance (not just a negative cash flow, but the ending balance is less than zero)? If so, then review the expenses and other cash outflows to decide what can be delayed. Can you delay the repayment of a loan? Can you delay the purchase of equipment? Can you purchase the equipment but defer payment by three months? Explore all strategies to prevent a negative cash position.

4. Look at the months where you have a positive cash flow (more money is projected to come in through sales than leave through expenses). The extra money coming in during those months needs to be set aside for the slower periods. For each month when more cash comes in than goes out (there's a positive cash flow), divide that amount of positive cash flow by the number of weeks in that month. You now have a weekly goal for each month. Some months you'll save more than others, depending on your expected level of sales.

    This method of saving is somewhat aggressive; any cash coming in above and beyond what you need for the budgeted cash outflows is saved. This helps you build a cushion for the unexpected expenses. Hopefully, at the end of the year, you will have more money saved than needed, and you can consider how to invest the money: whether in your business for the upcoming seasons or in your personal accounts.

5. At the end of the year, create a new cash flow budget for the coming year. Can you afford to take some dividends?

Once you've determined how much money to set aside each week/month, create a separate bank account where you will set it aside. If it's out of your primary checking account, you'll be less tempted to spend it.

3: Pay the Bills and Write the Checks

Now that you've determined which bills you will pay this week, go ahead: write the checks, record the transactions in QuickBooks and mail them (or pay online).

Every Month—Review, Refine and Stay on Target (1 hour)

Pick a time each month to sit down with your bank statements. Ideally, this will be in the first week of the month—after your bank statement and credit card statements arrive and before the credit card bill is due. While you can reconcile your bank statement and pay your credit card bill at different times, it will be more efficient to do them at the same time.

The month-end review gives you a chance to make sure things are on track, see how you're growing from the last year, check on profitability of different product lines and look at short-term cash flow planning.

1. Reconcile Bank Statements (including credit card bills)

a. Go to **Reconcile Bank Statement** from the Banking drop-down menu, or click on the **Reconcile** icon on the home screen.

b. On the first screen, select which bank account to reconcile, enter your ending balance as well as any interest earned and bank fees paid. When your credit card bill comes, reconcile it in the same way.

c. Check off all the cash inflows and outflows that are on your bank statement. If the bank statement shows an expense that's not in your QB register, be sure to add it.

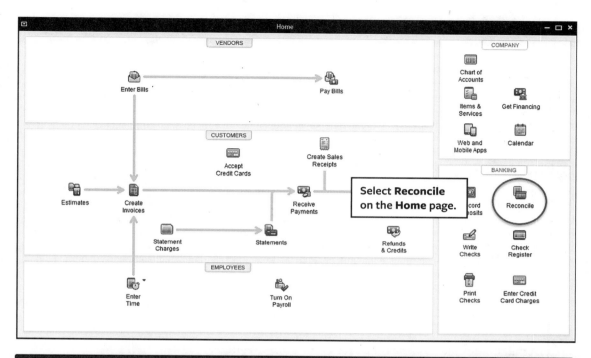

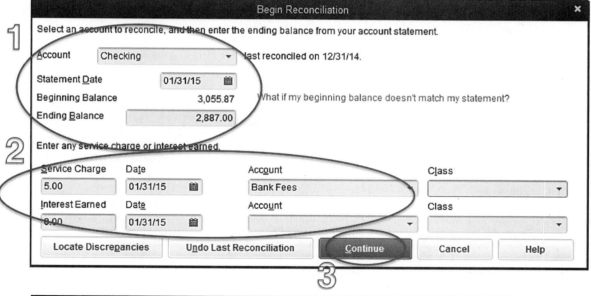

1. Select: the **Account** you are reconciling; the bank **Statement Date**; the **Ending Balance** in your account on the bank statement.

2. Enter any bank **Services Charges** or **Interest Earned** as indicated on the bank statement.

3. Select **Continue.**

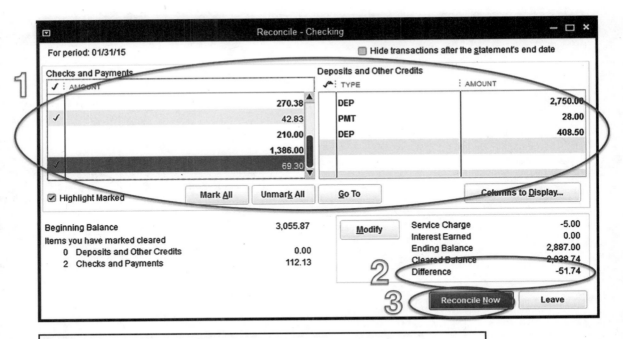

1. Select any **Checks** and **Payments** or **Deposits** and **Other Credits** that you have made but haven't yet hit your bank account.

2. The **Difference** should balance out to $0.00. If not, look for transactions that are on your statement but not in QuickBooks.

3. Select **Reconcile Now**.

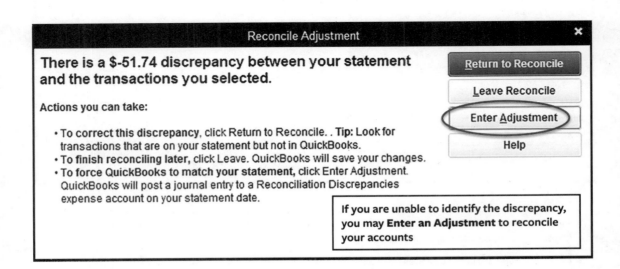

Managerial Thinking

d. Did you miss entering in a bunch of expenses? What adjustments to your systems do you need to make to better keep track?

e. Did you have any unexpected bank fees? What was the cause? What can you do to avoid them in the future?

2. Review the A/R Aging Summary Report

		A/R Aging Summary				— □ ✕

Customize Report	Share Template	Memorize	Print ▼	E-mail ▼	Excel ▼	Hide Header	Collapse	Refresh

Dates	Custom	▼	10/22/14 🗓	Interval (days) 30	Through (days past due) 90	Sort By Default	▼

10/22/15

Diversified Veggie Farm
A/R Aging Summary
As of October 22, 2014

	Current	1 - 30	31 - 60	61 - 90	> 90	TOTAL
Customer 187 ▶	0.00 ◀	0.00	100.00	0.00	0.00	100.00
Customer 202	0.00	0.00	0.00	0.00	359.00	359.00
Customer 221	0.00	0.00	0.00	0.00	120.00	120.00
Customer 222	0.00	0.00	0.00	0.00	35.00	35.00
Customer 283	0.00	0.00	0.00	0.00	170.00	170.00
Restaurant 3	72.00	342.00	464.00	336.50	493.50	1,708.00
Customer 333	0.00	730.50	1,127.00	721.75	754.00	3,333.25
Restaurant 1	0.00	0.00	0.00	0.00	32.00	32.00
Customer 371	0.00	0.00	0.00	0.00	32.50	32.50
Customer 392	0.00	0.00	0.00	0.00	-476.00	-476.00
Pizzeria	0.00	0.00	0.00	0.00	-72.00	-72.00
TOTAL	72.00	1,072.50	1,691.00	1,058.25	1,448.00	5,341.75

a. Go to **Reports > Customers and Receivables > A/R Aging Summary**.

b. Review which customers owe you money. Customers that are in the **current** column have outstanding invoices but are not late in paying. Customers in the 1–30 columns and beyond are late in paying.

c. If customers have several outstanding invoices, create a **monthly statement** and email.

Managerial Thinking

a. Are customers paying you on time? If not, what do you need to do to ensure they do? Two options:

- "Fire" clients that don't pay on time.
- Assess finance charges to chronically late payers.

3. Look at the Profit and Loss Statement

a. Go to **Reports > Company and Financials > Profit and Loss Standard**.

b. Run a profit and loss statement for the previous month. Click the **customize** button (on the Mac, click on the **Options** icon), and add a column for **same period previous year.**
- How does revenue this year compare to last year? Is it increasing? If not, can you explain why not?
- Are profits greater than last year?

c. Run a second profit and loss statement for the year to date, click the **customize** button; add a column for **class.**
- Which class is most profitable? What can you do to increase those sales?
- Look at the expenses of the different classes. What can you do to decrease them?
- Are you losing money in one class? What can you do to make it profitable?

4. Review your Balance Sheet

a. Go to **Reports > Company and Financials > Balance Sheet Standard.**
b. Run a balance sheet for the previous month. Customize the report and add a column for **same period previous year.**
- Compare the total assets from last year to this year.
- Compare the total liabilities from last year to this year.
- Compare your accounts payable to accounts receivable.

Managerial Thinking

a. Are you assets increasing over last year? Are your liabilities decreasing? If you want your business to grow every year, then assets should be increasing to keep pace and the liabilities should be decreasing. Are you building (increasing) equity?
b. Accounts Payable shows money that you owe vendors. Does your cash balance look like enough to cover the upcoming bills?

Compare to Budget/Cash Flow

a. Compare revenue projections to actuals. Are they better than you anticipated? Congratulations!!! If not, what happened? It's easy to blame external forces such as weather or a weak marketplace, but stretch yourself to consider factors that you can control.

- What could you have done to improve production during unfavorable weather conditions?
- Can you be less risky in your plantings?
- How could you modify your sales approach to weather a weak economy?

Year End—(1 to 2 hours)

Whether you celebrate the holidays or not, the time around Christmas and New Year's is joyfully quiet. Around January 5, well after the hullabaloo of Christmas has faded and you've recovered from New Year's celebrations, I suggest sitting down with your books. For all reports, be sure to select the date range of January 1–December 31 of the year in review.

1. Run a Profit and Loss Standard Report

a. Customize the report to **display columns by month** across the top, and add a subcolumn for **percentage of income**. Some questions to ask yourself:
- Did you earn as much money in each month as you had hoped and planned for?
- In which month did you have a negative net income? Is that what you expected? What would you do differently?
- How did your sales mix change throughout the year?
- In looking at the year as a whole (the last column), do any expenses jump out at you as being too high? Look at both the dollar amount and the percentages. Pay particular attention to expenses that are above 5% of total revenue. Do you have an explanation for why they are high? What can you do to bring them down?

2. Run a Profit and Loss Statement Standard

a. Customize the report to **display columns by class** across the top, and add a subcolumn **percentage of income**. Some questions to ask yourself:
- Which was your most profitable **class** as a percentage? As a dollar amount?
- What are there opportunities to grow the more profitable areas? How much can you increase production and not flood the market?
- Which was the least profitable **class**? What can you do to improve its profitability? Must you retain these products to stay attractive to your customers?

3. Run a Balance Sheet Standard

   a. Customize the report to **Add subcolumns for previous year.**
      - Did your equity increase over the prevous year?
      - Compare total equity to total liabilities. Is equity greater?
      - If liabilities are greater, what can you do to manage and reduce them?

4. Create a Cash Flow Budget for the Next Year

   a. Review your income statement from last year. Create a Profit and Loss Standard for the previous year. Click **Modify Report**. Select **Display Columns by Month** across the top. Also check the box for **% of income**.
   b. Consider how each number will change for the coming year. Do you expect your sales numbers to increase? If so, by how much? Do you expect your expenses to increase? Will they increase as a percentage of sales or as a fixed percentage?
   c. What big purchases do you want to make in the coming year? Will you need to take out a loan or otherwise get financing?
   d. Use the Quick and Dirty Cash Flow template (juliashanks.com/TheFarmersOfficeTemplates/) to project the coming year.

QuickBooks Quick Tips

The best advice I can give you about QuickBooks is to be consistent. Be consistent in the habit of entering in your receipts, bills and deposits. Be consistent in looking at your budget and comparing it to actuals. And be consistent in how you record things (seeds should always be entered as *seeds*).

Here are seven quick tips to make the most out of your QuickBooks:

1. **Separate your revenue** into major categories. For a farm, this may mean farm stand, farmers markets, CSA and wholesale.
2. **Separate your cost of goods sold** into the same categories as your revenues.
3. **Use classes** to track expenses associated with specific products such as produce or flowers.
4. **Treat farmers markets as a customer.** Use **customer: jobs** to track sales and expenses from each market you attend, in addition to using the **customer: jobs** to track actual wholesale and other customers.
5. If you have regular payments, such as for a bank loan or rent, click on the **Memorize** button so that QuickBooks automatically enters it.
6. **Use the memo line** to remind yourself what different transactions were

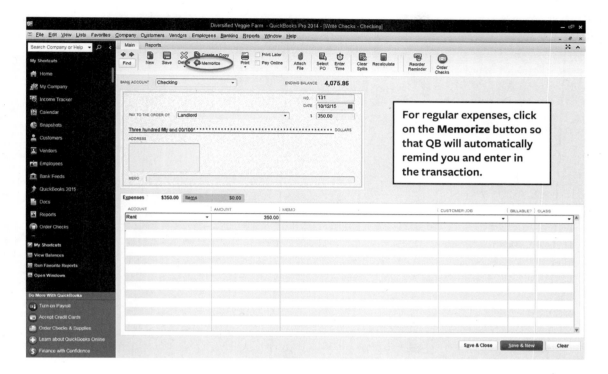

For regular expenses, click on the **Memorize** button so that QB will automatically remind you and enter in the transaction.

about. When you look back on your finances at the end of the year, you may wonder what a seemingly random check was about. Maybe it was $500 for supplies. If you make a note in the memo line, it will be easier to recall.

7. **Don't use a Miscellaneous category.** It's too easy to dump things into that category rather than figuring out the best way to classify a transaction. With too many transactions there, you lose the richness of sorting expenses and purchases by category.

Notes

1. If you go straight to **record deposit** without first **receiving the payment**, QB won't know to match the check with the invoice.
2. Remember, an asset is a type of account on the Balance Sheet. It is a future benefit as the result of a transaction. The future benefit is money in your bank account; the past transaction is the sale.
3. If you did not already create this account, you can set it up now by opening the **Chart of Accounts** window, and creating a new account. The account type is **Other Current Asset**. When you are in the account setup window, you will find **Other Current Asset** in the pull-down menu for **Other Account Types**.

CHAPTER 8

Digging into the Numbers
and Beyond

We're just getting into the good stuff! All that bookkeeping and cash flow planning is a long preamble to get you to the real fun—managerial accounting: using the numbers to make decisions about how to improve and grow your business. Having a handle on your financial reality is a big part of day-to-day operations. And once you understand the numbers, you can hone your other financial and business management skills.

In this chapter, we'll discuss how to read, interpret and manipulate the numbers to give you meaningful information about your business and the tools to make decisions.

Cost Accounting

How much does it cost to produce a skein of yarn, a bouquet of flowers, a gallon of milk or a case of tomatoes? If you knew that you were losing money on one of them, would you still grow them? Probably not.

Most farmers just look at the bottom line: "Did I make more money than I spent?" If so, then they don't make changes. The only time they might investigate their financials is in years they are not profitable. But you're not "most farmers." You want to know more—you want to make money on each and every product; and ensure you're pricing correctly. You know that you can improve profits by working smarter (understanding your costs and pricing correctly) and not harder (just growing more).

By understanding which products are more profitable, you can hone your operations. If you earn $2 a case for tomatoes and $6 a case for mesclun, then wouldn't you want to grow more mesclun? Would you reconsider your pricing structure? Brett from Even' Star Organic Farm learned this when he analyzed his crops; and it informed his decision to further diversify his crops *and* to raise his prices.

Brassa Farm in Marion, MA, did a similar analysis.

Pricing Pigs

Brassa Farm had been raising feeder piglets for the last 10 years. Their customers were farmers who bought them for about $95 each and then raised them for slaughter (and meat). In addition to the piglets, Brassa grew vegetables on a few acres and sold the produce at their farm store. Business was brisk, but they just weren't getting ahead.

When we sat down to look at the numbers, they explained that the piglets were relatively cheap to raise. Each piglet had $10 in health costs plus about $14 for feed for the 8 weeks until they were sold. Brassa figured they earned $71 per piglet.

"What about the boars and the sows?" I asked. "How much does it cost to maintain them? Isn't breeding part of the cost of raising the piglets?" They looked at me quizzically. In all their calculations, they never considered the cost of the boars and sows that "generate" the piglets. We calculated the cost to maintain their 2 boars and 10 sows. With the bedding, supplies and feed, the total came to almost $19,000 per year. Those 12 animals produced 200 piglets per year, which meant the cost of "generating" a single piglet was $95! Added to the cost of raising the piglet to 8 weeks, the real cost was $119. The Brassas were losing $24 per piglet.

It was clear this was not a sustainable business model. If they wanted to improve their profitability, they needed to rethink the pig business. They decided instead to raise the piglets to full size themselves and sell the meat in their farm store. Three years later, revenue had increased tenfold, as did their profits.

You may ask, "How could they not see they were losing money on piglets?" Simply put, they looked at their business as one unit—and the profits from their vegetable and farm store operations masked the losses from the piglets. Taking an informed look at their business and operations forced the Brassas to challenge their base assumptions about their business and make necessary changes to improve profitability.

Marketing vs. Operating Expenses

Sunny and Marshall opened Gilmanton Winery in 2010 in New Hampshire. Over the course of four years, they built a tasting room for retail wine sales and a restaurant, and established a wholesale customer base. As they approached their mid-fifties, they wanted to build a business that would support them in their retirement. In addition to the wines, Sunny had a passion for knitting. They bought several alpacas—primarily as a source of wool for Sunny, and also as an added attraction to the winery. Sunny sold skeins of yarn, as well as knitted scarves and hats in their winery retail store.

When I toured the winery, Marshall proudly showed off his wife's scarves and hats, commenting on the hard work required, up to 8 hours, to knit these beautiful pieces. His mention of 8 hours set off red flags in my head. At $50 per scarf, Sunny was earning $6.25/hour, assuming there were no other costs. I couldn't see how this would be profitable. We set up Sunny and Marshall in Quick-Books, and Sunny began tracking all of her sales and expenses, and sorting them by class. She separated all transactions by the different revenue streams—her alpacas, the restaurant at the winery and the wines themselves. After three months of tracking, she had some good data. It was with little surprise that when we looked at the profitability of the alpacas—with the expenses of feed, shearing and processing the wool—she was losing money (and this didn't even factor in her time).

Armed with hard numbers, Sunny had an important decision to make: should she discontinue raising alpacas, raise her prices on the scarfs or classify the alpacas as a marketing expense since they draw people to the winery?

Good accounting clarifies options. For Sunny, she may not be able to raise the prices for her scarves (because they won't sell if they're too expensive), but she can maintain the alpacas as a passion or as a marketing tool. And as she thinks about growing her business, she needs to properly frame the alpacas: not as a revenue stream but as a marketing expense that happens to provide a little income.

Beyond the alpacas, Sunny and Marshall needed to grow the business so it would generate enough profits to support them in retirement. It had already become clear that the alpacas were not the ticket. The winery and restaurant on the other hand were making money. Is the restaurant sufficiently profitable to try and grow? Is it better to sell wholesale or through their retail winery? By calculating the true costs of each sales channel, they could make an informed decision.

Understanding the true cost of production, gives you information to:
- set prices
- decide where to focus your energies
- understand the most profitable growth opportunities.

It's tempting to try these calculations before you've launched your business. And, indeed, it can provide a good starting point for your initial season in thinking about prices. However, this type of analysis yields better insight when conducted at the end of a season—when you have actual numbers to look at. You can then use this information to make adjustments for upcoming years.

Calculating True Costs

As you no doubt experienced with setting up and managing your transactions in QuickBooks, it can be difficult to tease out your profitability on each item you sell. You buy feed for many animals and seeds for many vegetables. The black plastic covers many beds. All pigs may get the same feed, but some are destined for meat and others are sold as feeders. How do you isolate the costs just for yarn or milk or tomatoes or pork?

Cost accounting is further complicated in that not all units are the same. Some things you think about in acres (or rows) and some things you think about by the case; animals are raised by the "unit" and sold by the pound. Mesclun is planted by the row, but harvested by the pound. So you need to understand what you're measuring and how to convert the numbers for the comparable unit of measurement.

We will work through a few examples of cost accounting. In these examples, we will assume that each hour of labor costs $15 ($13.00 per hour + $2.00 for payroll taxes and worker's comp).[1] When you do this type of number crunching on your business, you may assume labor is more or less. But whatever you do, even if you do all the work yourself, include labor. First, in order for your business to be sustainable, you need to be able to pay yourself. Second, if you're not doing the work, then you need to pay someone else. By including all labor in your cost analysis, you will have a better sense of the true cost of running your business.

Further, we will make many other assumptions that are too great to list—what varieties of each crop do you grow, how far is the farm from the market, is the farm using organic or conventional growing practices, does it have chickens to "generate" compost, and so on. Some of these numbers may make sense for your operation, others may not. I caution you against taking the following numbers as hard truths. They are for illustrative purposes only.

Example—The Cost of Production: Tomatoes[2]

The costs associated with a case of tomatoes are buried in many different places: some numbers you'll find in QuickBooks, some will be on employee time cards, and some you'll just know. You may need to extrapolate some numbers as well. For example, a fully planted acre of tomatoes requires 8,500 plants. A seed tray has 30 cells. Therefore, you will need about 285 seed trays for enough tomato plants for one acre.

For now, we will focus just on the direct costs. There are also costs associated with overhead, utilities and the equipment you use to prepare the beds, plant and harvest. We'll address these costs a little farther along.[3]

1. Figure out the conversion factor.

A make-believe farm in West Virginia plants an acre of tomatoes, which yields about 25,000 pounds. They pack 20 pounds of tomatoes into a case. Therefore, an acre yields 1,250 cases of tomatoes (25,000/20). In order to convert acres to cases, divide by 1,250.

Where to find the numbers:

- Keep track of how many cases of tomatoes you yield in a season. You can do this by using **items** in QuickBooks (see page 97) to track your sales numbers, or you can write down yields on a clipboard in the barn as you bring each case of tomatoes in from the field.

- Ask other farmers in your region what their yields are. Consider their growing practices and soil conditions, and how that may impact a comparison to your potential yields

2. Calculate the cost to get the tomatoes into the ground.

Labor (per acre, per season):

- The seeds need to be started in the greenhouse, and it takes about 4 hours to fill the trays with soil and seed. ($60)
- Over the course of 7 weeks between starting the seeds and transplanting them into the fields, you'll water every day, transplant the starts into celled trays and harden them off. This will take a total of 82 hours. ($1,230)
- Once the seedlings are ready to transplant, it takes about 300 hours to fertilize the soil, lay plastic mulch, set up the irrigation system and the stakes and get them all in the ground. ($4,500)
- After that, there's the field tending: tying up the tomato plants to the stakes and weeding, 342 hours. ($5,130)

Total Labor Cost: $10,920

Where to find the numbers:

- Keep a log in the greenhouse/barn where you can jot numbers down.
- Ask employees to track their hours on different crops and on different tasks.
- Some tasks, like filling a seed tray with soil and seeds, are completed with multiple crops at one time; a little extrapolation is needed. How long does it take to fill one tray, and how many trays do you need? Do similar extrapolations with other tasks.

Supplies (per acre, per season)

- Potting mix: $660
- Seed trays: $135
- Seeds: $136
- Black plastic mulch: $650
- Amendments (including cover crops and fertilizer): $375
- Stakes and ties: $4,128

Total Supply Costs: $6,084

Where to find the numbers:
- If you are tracking cost by crops (using jobs or classes) in QuickBooks, you can run a Profit and Loss report and add a column for jobs or classes.
- If you are track costs in QuickBooks more generally, then you will need to dig deeper: look at the invoice from your seed company—what did you spend on tomato seeds? Look at the invoice for the black plastic—how much does it cost per foot? How many feet did you use for your planting?

Total cost per acre (labor and supplies): $17,004
- Cost per case: $13.60
- Cost per pound: $.68

3. Calculate the cost to get the tomatoes out of the ground.

We calculated the cost of getting tomatoes into the ground by the acre. Now that we're calculating the cost to get tomatoes out of the ground, we're thinking in terms of cases (or pounds).

Labor (per case)
- Harvesting: 15 minutes per case
- Sorting/Packaging: 15 minutes per case
- Total labor per case: $7.50

Supplies (per case)
- $1.00 for cardboard pint containers for 20 pounds

Total cost per case (labor and supplies): $8.50
Total cost per pound: $.43

4. Calculate Total Cost: Seed to Case
- Into the ground: $13.60 per case
- Out of the ground: $8.50 per case
- Total cost: $22.10 per case or $1.11 per pound

Certainly, you don't want to sell any crops for less than their cost of production. Beyond that, you need to cover the costs to bring them to market and overhead.

Example: Bringing Product to Market

Whether you sell your tomatoes wholesale or retail, the cost of production is more or less the same. As you layer in the cost of selling, you gain an understanding of the volume of sales necessary to make a sales channel worthwhile.

Let's say you sell the tomatoes wholesale. Consider all the costs—both in labor and hard costs—you incur to sell your product. This includes calling your customers, loading the truck, and making the delivery.

Labor
- Call with customer to get order (10 minutes)
- Load the truck with product for each customer (10 minutes)
- Delivery time, driving to and from farm (40 minutes, plus 30 minutes for each additional drop-off)
- Total time: 1 hour, $15 per delivery.

Other expenses
- Fuel $5 per drop-off

Total cost $20 per delivery (not including the cost of maintaining the delivery truck)

The cost of each delivery is a fixed cost; meaning, whether you deliver one case or 20 cases to a customer, it will, more or less, be the same. In order to improve profitability, you must deliver in volume. You can see the difference in the tables below. This example assumes a sales price of $40 per case.

TABLE 8.1. Profitability of selling one case of tomatoes

Sale Price	$40.00
Cost of Production	$22.10
Delivery	$20.00
Gross Profit (Loss)	–$2.10

TABLE 8.2. Profitability of selling 10 cases of tomatoes

Sale Price	$400.00
Cost of Production	$221.00
Delivery	$ 20.00
Gross Profit	$159.00

It's tempting to oblige a customer's request for a delivery of $50 worth of produce. After all, $50 is money you wouldn't otherwise have in your wallet. This analysis, however, demonstrates the importance of setting order minimums for your wholesale customers. If the cost of each delivery is $20, then you must have a gross profit (not revenue, but the profit after accounting for the cost of production) of at least $20 just to reach break-even. Beyond that, you want additional profits to cover overhead expenses.

Of course, you don't just sell tomatoes. You sell cucumbers, mesclun, eggs and many more different products. To really understand how to set a minimum, calculate an average cost of production as a percentage of the sale price. In other words, on average, for every dollar of produce sold, what does it cost you to grow it?

If you calculate the cost of production (COP) for each of your crops as we did above for the tomatoes, you can then take an average. Let's use some sample numbers:

TABLE 8.3. Cost of Production per Case

	COP/Case	Wholesale Sales Price	COP %
Tomatoes	$22	$40	55%
Cucumbers	$12	$26	46%
Early Glow Strawberries	$54	$72	75%
Mesclun	$19	$28	68%
Eggs	$65	$75	87%
Kale	$12	$26	46%
Radishes	$16	$22	73%
Average Cost of Production			64%

The average cost of production is 64%.

Doing this sort of analysis on all your crops and products is tedious. I wouldn't expect or even suggest that you do it. Nonetheless, I recommend that you do this analysis with at least three to five of your crops or products—focusing on the products that you sell the most of or your gut tells you are the least profitable. Once you do it a few times, you'll develop your intuition of the numbers, and you'll be able to do future analysis in your head with a dose of gut instinct.

Even if you don't look at individual crops, you can look at total numbers. This calculation is easier though doesn't provide the detail that the above analysis does. Look through your expenses and pull out all costs associated with production, including labor,[4] supplies, seeds, soil amendments and mulch. Add up all those costs and divide it by total sales. This is the average cost of production (ACOP).

$$\text{Average Cost of Production} = \frac{\text{Total Production Costs}}{\text{Total Sales Revenue}}$$

For the following examples, we will assume an ACOP of 64%.
    Therefore, the average gross margin is 36%.

$$100\% - 64\% = 36\%$$

100% of Sales minus COP equals Gross Margin. Also assume for the following example that the average delivery cost is $20.

Break-Even Analysis

We know that we want the revenue from each customer when multiplied by 36% (the gross margin) to be greater than $20 (the cost of delivery)—that is the point at which you break even; the revenue you earn covers the cost of production. As a formula, it would be expressed like this:

$$\text{Break-even Sales Amount} \times 36\% = \$20$$

In order to calculate the minimum sales amount, we divide both sides by 36%

$$\text{Break-Even Sales Amount} = \frac{\$20}{36\%} = \$55.56$$

Of course, your numbers will be different from the above example. To calculate your break-even order volume for wholesale customers use this formula:

$$\text{Break-Even Sales Amount} = \frac{\text{Delivery Cost}}{\text{Average Gross Margin}}$$

Where Gross Margin is calculated by:

$$\frac{(\text{Sale Price} - \text{Cost of Production})}{\text{Sale Price}}$$

The gross margin can be an average of many different crops.

Farmers Markets[5]

We can do a similar analysis with the cost of selling at the farmers market.

- **Labor:** For every market day, you log about 12 hours in labor. This includes the time it takes for you and your crew to pack your truck, drive to the market and back, set up your stand and break it down, as well as the actual time you sell during market hours.
  - Total Labor Cost: $180
- **Hard Costs:** In addition, you pay a market fee. You purchase tents, tables and displays and bags and cardboard containers for customers to take away their goods.
  - We can estimate the total cost of the tent, tables and displays at $1,500. These supplies generally last for 2 year. For a 20-week season, the cost is $37.50/week ($1,500 divided by 2 years divided by 20 weeks).[6]
  - The market fee is $25/week.
  - Fuel to get to market: $30
- **Total Fixed Costs for going to the farmers market each week: $272.50**
- There's an additional variable cost of the packaging. The packaging (shopping bags and cardboard boxes) is 1.6% of total sales.

Presumably, the profitability of your crops is better at the farmers market than wholesale as you sell at a higher price.

Table 8.4. Sample Financials and Spreadsheets—Cost of Production (per case)

| | | Farmers Markets | |
Case	COP/Case	Sales Price	Gross Margin
Tomatoes	$22	$50	56%
Cucumbers	$12	$33	64%
Early Glow Strawberries	$54	$90	40%
Mesclun	$19	$35	46%
Eggs	$65	$94	31%
Kale	$12	$33	64%
Radishes	$16	$28	43%
Average Cost of Production			48.9%
Packaging			1.6%
Total Cost of Production + Packaging			50.5%
Gross Margin			49.5%

The average cost of production is 48.9% of the sales price and packaging is 1.6%. The total variable costs are 50.5%. Therefore, the gross margin is 49.5% (100% − 50.5%). In order to cover the cost of going to the farmers market, in this example, you'd need to sell at least $550.51 worth of product.

Cost of Selling at Market = Break-Even Sales Volume × Average Gross Margin

$$\text{Break-Even Sales Volume} = \frac{\text{Cost of Selling at the Market}}{\text{Average Gross Margin}} = \frac{\$272.50}{49.5\%} = \$550.51$$

The good news is that the true cost of growing your product already includes your labor. The profits above and beyond the $550.51 in sales will be applied to your mortgage/rent, insurance, maintenance on the farm and so on.

Entrepreneurial Thinking

As you evaluate the profitability of farmers markets and different wholesale customers, you will consider many factors. Understanding where you make the most money will inform these decisions.

Taking Averages

When you run the numbers in this way, consider what kind of year you're having. Are you rocking the tomatoes, or has it been unusually cold and wet. A great year will likely show more profitability than a lousy weather year; beware of basing your numbers on the best year. As you apply this analysis to future years, you may want to adjust the numbers accordingly. You can never predict the weather before it happens.

Allocating Overhead

The above analysis factors in the *direct* expenses of your different products, and is a great tool for determining selling prices and sales minimums. To be sure, there are many other expenses involved in managing your farm, both direct and indirect. The tractor, for example, can prepare your tomato beds, cucumber beds as well as your chicken pasture all in the same day. How do you allocate the cost of fueling, maintaining and depreciating your tractor to each product? In addition, you pay for rent or a mortgage on the land, utilities, the accountant to file your taxes, office supplies and so on.

By taking the extra step to allocate the other expenses, you gain insight into whether the products you grow provide enough profit to the business to make them worthwhile.

All of your expenses can be divided into three categories:[7]

- **Direct Operating:** Expenses that can be directly attributed to a product, including feed for your chickens, stakes for your tomatoes or bottles for your milk.
- **General Operating:** Expenses that are directly related to the operation of your business but not easily attributed to a single product or sales channel, including the cost of running your tractor and small tools.
- **Overhead:** expenses that result from running a business, including insurance, tax preparation expense, advertising, and office supplies.

The direct operating expenses we allocated earlier. The other two categories can be allocated proportionally by revenue or acres (or square footage) in production.

Allocation by Production

You can use the percentage of acres in production as a method to allocate overhead expenses. How much space do you allocate for tomatoes, chickens, flowers or other crops?

Let's say you have five acres in production and $18,000 in general operating expenses.

Crop	Acres in Production (Total 5 acres)	Acres in Production as a percentage of Total Acreage	Allocation of $18,000 in general expense
Tomatoes	1.25	25%	$4,500
Meat Birds	0.75	15%	$2,700
Layers	0.75	15%	$2,700
Greens and Greenhouse	1.25	25%	$4,500
Flowers	1	20%	$3,600
Totals	5 acres	100%	$18,000

1.25 acres are used for tomatoes; 25% of the total acres in production is used for tomatoes:

$$\frac{1.25}{5} = 25\%$$

Therefore, 25% of general operating expenses is allocated to tomatoes. If general operating expenses totaled $18,000, then $4,500 of general operating expenses are allocated to tomato production.

Allocation by Revenue

Alternatively, you can allocate overhead based on the percentage of revenue each product contributes to your business.

Let's say the farm earned $135,000 last year, and total overhead expenses for the year were $12,000.

Crop	Sales of each crop/product	Percentage of Total Revenue	Allocation of $12,000 in overhead expense[8]
Tomatoes	$28,000	21%	$2,489
Meat Birds	$37,000	27%	$3,289
Layers	$32,000	24%	$2,844
Greens and Greenhouse	$20,000	15%	$1,778
Flowers	$18,000	13%	$1,600
Totals	$135,000	100%	$12,000

To calculate the distribution of overhead expense of tomatoes to the farm's bottom line, allocate 21% of overhead expenses to the cost of producing tomatoes.

$$21\% \text{ of } \$12,000 = \$2,489$$

In the earlier example, $4,500 of general operating expenses were allocated towards tomatoes.

Total allocation of expenses for tomatoes is:
- general production costs: $4,500
- overhead: $2,489

For simplicity, we'll use the earlier calculations that the average cost of production of tomatoes is 55%. If you sold $28,000 worth of tomatoes, you can estimate the production cost of $15,400.

Therefore, the tomatoes generated $7,611 towards the farm's profitability.

$$\$28,000 - \$4,500 - \$2,489 - \$15,400 = \$7,611$$

Is it worth your effort/time to grow tomatoes? That's for you to decide!

—— Which to Choose: Allocation by Production or Revenue ——

You can choose your method based on how you think expenses are most impacted. General operating expenses, such as the cost of running your tractor, is more a function of acres in production. The cost of insurance is likely a function of revenue. You can use just one method for all expenses that are not directly related to a revenue stream, or you can use both methods and further categorize your expenses by how they are impacted. Using just one method is simpler. However, you may be seeing the power of numbers and want to take the extra step. Using both methods is more complicated but yields more accurate numbers.

As you consider the profitability versus effort of various crops, animals or products, you may be on the fence about a few. Do you discontinue those crops, raise prices or figure out a way to reduce expenses?

Here are some questions to ask yourself as you weigh the decisions:

- What other crops are you growing? How profitable are they? Will they contribute sufficiently to your bottom line?
- How would you make up the revenue from a discontinued product? Would you need to make it up?
- Do you rely on one crop/product to draw customers to purchase other products?
- Can you raise your prices? If you do, how will your customers respond? Do you need to explain the value of your tomatoes over the competition that may sell them for less?

Good stuff! But you may be wondering, "How can I possibly track the sales of my different products? The farmers market is way too hectic to track how many pounds of tomatoes I sold, or any other crop, for that matter. Not to mention, my farm store is on the honor system. How can I know how much I'm selling of what?"

Inventory is how.

Managing Inventory

For many situations, a good point-of-sale (POS) system can resolve the issues of tracking sales by product or category; even a system as simple as Square. Counting inventory also works as a way to track sales without any special equipment or software. Tracking inventory has the added benefit in that you can ensure that product isn't stolen and is properly rotated.

Managing Inventory

Reilly and Jack are meat farmers in Hudson Valley, NY, selling their meats at their own farm store as well as to restaurants in Manhattan. They were on the verge of purchasing new land, and wanted to know which animals were most profitable so they could know the best opportunity to expand production. They worked diligently to get their QuickBooks in good shape so they could understand the true cost of each type of animal they raised. They tracked all their expenses by animal and were able to allocate the overhead appropriately. In terms of understanding their profitability, they needed to know exactly how much they earned from each type of animal. They knew exactly what they sold to the restaurants as they created invoices for each sale. The farmers market and farm store sales were more difficult to untangle. The farmers market is such a hustle, there is barely a second to breathe, much less track each sale. If they stopped to write everything down, they'd lose customers who didn't want to wait; a far worse state then not knowing exactly how much pork vs. beef vs. chicken they were selling. In the farm store, they received payment with the honor system. They thought about leaving a clipboard for the customers to write down what they took, but they didn't trust the customers to write down the correct item.

They needed to create a system to track inventory so that they could track sales.

Consider this equation:

$$\text{Ending Inventory} = \text{Beginning Inventory} + \text{Restocks} - \text{Product Sold}$$

Beginning Inventory is what you have at the beginning of the day or week in your farm store. It's what you take with you to the farmers market. If you have a farm store, you may *Restock* the shelves, refrigerator and freezer throughout the week with new product. And through the course of the week or market, you'll sell product (*Product Sold*). Beginning Inventory, Restocks, Product Sold, Ending Inventory: you can think of these things in dollar value or as units. This could be 10 pounds of bratwurst, 30 pounds of ground pork, 5 cases of tomatoes, 6 dozen eggs. Or you could count it as $200 worth of pork, $125 of tomatoes and $36 worth of eggs. For simplicity sake, I recommend thinking in dollar values.

If you turn the equation around, you have the formula to calculate the amount of product sold.

Product Sold = Beginning Inventory + Restocks − Ending Inventory

If you count your inventory at the beginning of the week (or farmers market) and again at the end of the week (or farmers market), and track any restocks,[9] you can easily calculate what you sell. By taking this extra step, you can now track both the revenue by sales channel (farmers markets vs. farm store) as well as revenue by product (produce vs. pork vs. poultry). This additional data point enables you to examine the profitability of each product line. You already track the expenses by product and sales channel; now you can do the same for revenue.[10]

Calculating Sales by Counting Inventory

Table 8.5 shows an example of how Reilly and Jack tracked inventory in their farm store. They counted how much product was in the freezer and fridge at the beginning of the week, how much they restocked on Tuesdays and Thursdays, and calculated the dollar values. At the end of the week, they counted again, figuring out the dollar value of what was left on the shelves and in the freezer. This showed them how much product was sold.

Other Benefits of Tracking Inventory

By using this method of counting inventory, you have a checkpoint for how much money you should be earning. In tracking how much you sold, you also know how much you money you should have received.

TABLE 8.5. Farm Store Inventory Worksheet (Week of November 5)

Count	Price/Unit	Beginning Inventory	Restocking							Ending Inventory
			Sun	Mon	Tue	Wed	Thu	Fri	Sat	
Beef (dollar value)		$100.00			$125.00		$150.00			$90.00
Pork (dollar value)		$250.00			$75.00		$50.00			$240.00
Chicken (dollar value)		$208.00			—		—			$26.00
Milk (½ gallon bottles)	$9.00	4			6		4			1
Eggs (dozen)	$6.00	14			14		10			2
Tomatoes (pounds)	$4.00	20			15		15			0

Money in Cash Box Week-End	1,150.00
Value Taken	1,135.00
Cash Box Was Overpaid	$15.00

	Total Sold for QB
Beef	$285.00
Pork	$135.00
Chicken	$182.00
Milk	$117.00
Eggs	$216.00
Tomatoes	$200.00

At the end of the farmers market, you count the money in your cash box, and you count your ending inventory. Based on your inventory, you calculate what you sold.

If the money in the cash box doesn't match the estimated revenue based on your inventory, then you could have a problem. It could be as simple as not tallying customer orders correctly, or giving incorrect change. It could also be something bigger, including:

- An employee is stealing from you.
- A customer is underpaying and/or stealing from you.
- Too much product is being thrown away.

Some weeks you may be short, some weeks you may have more cash than you should. If you notice that your cash box is consistently short, then you should investigate to find the cause.

In addition, the physical act of counting inventory helps ensure proper rotation; that case of cucumbers that was shoved in the back of the cooler is likely to be unearthed (and sold instead of fed to the chickens).

See Inventory Tracking template and instructions at: juliashanks.com/TheFarmersOfficeTemplates/

Marketing and Pricing Strategies

When you initially thought about how to price your goods, perhaps you looked at what other farmers were changing and just copied them. Or maybe, you wanted to keep your prices in line with the local supermarket. After all this cost-accounting analysis, you may realize you will need to raise your prices if you want to be financially sustainable. To determine the best price, look at both sides—what is the competition charging, and what does it cost you to produce your goods.

I don't need to tell you that increasing your prices is not as simple as it sounds. Will your customers still buy from you if you raise your prices? Does raising your prices misalign with your mission of providing local sustainable food at an affordable price? On the other hand, if you charge less than the competition, there may be room to raise your prices and remain competitive.

Most of your customers do not shop on price alone. If they do, then they are not long-term clients and will drop you as soon as they find someone else selling at a lower price. You do not want to get into a price war; you will lose. They buy from you because they get value from you: better quality produce, customer service, the halo effect from supporting local farmers.

Communicating Value

In order to command a higher price, you need to explain to your customers why your products are priced as they are. You can communicate this through your packaging, your signage, your emails and your sales pitch. The message you communicate depends on what your customers value about you and the products you sell.

I want to clarify one point—what *you* think are the key features of your products may not be what your customers value. You may think your heirloom varieties are the most important, but your customers may care more that you grow organically. If you are not sure about the value you provide your customers, ask them. Specifically, ask regular customers with whom you've developed a relationship and you trust will provide honest feedback.

Here are some examples of messages that communicate value to your customers:

- My product is fresher than what is available through the supermarket or wholesale distributor. It will last longer, and in the end, the customer will save money.
- It tastes better than anything else available.

- I grow heirloom varieties not commonly available, and they have superior flavor.
- I grow using organic practices.
- I provide recipes and cooking tips with the weekly CSA shares.
- The CSAs are dropped off at convenient locations.
- I offer pick-your-own (PYO) to my CSA members.

Further, consider what you can do to maintain and deepen relationships with existing customers. What can you do to "wow" customers to stay memorable? What are some low-cost, high-impact strategies you can use? Remember, it is cheaper to maintain existing customers than to attract new ones. This could be as simple as offering small samples at your farm stand or creating dynamic displays.

Loss Leaders and Product Mix

After careful consideration, you may decide that, in fact, you cannot raise your prices, or you can't raise them enough to ensure their profitability. To be clear, it's okay to have one or two products that don't carry their weight,[11] but the products that don't generate profits need to provide another value to the business. For Sunny at Gilmanton Winery, the alpacas draw customers to the winery. Once there, they visit the wine store or restaurant and spend money. For Brett at Even' Star Farm, he uses the tomatoes to attract customers. Once he establishes the relationship, he can sell his other (more profitable) products.

These items are considered loss leaders. Most businesses have one or two. The key is to make sure they are not your top-selling items, and you have other items with sufficient profitability to make up for them. If your loss leader is your top-selling item, then consider capping its production or stop selling it altogether. Continuing to sell unprofitable items at a high volume will drive you out of business.

Benchmarking

Benchmarks, in general, are a standard by which performance is measured. For a farm business, they can help you evaluate the effectiveness of moderating your expenses and growing sales. They establish a target of where your expenses should be, and how fast you want to grow. You can set your benchmarks to industry best practices or to a customized goal that is suitable for your business. Not all farm businesses fall into the same mold; only you can judge what is appropriate for your business.

I've seen a few examples of benchmarks for small, diversified farms:

- Labor Cost: 35% of sales
- Operating Expenses: 40%–50% of sales
- Seeds and seedlings: 3% of sales (included in the direct operating expenses)
- Net Profit: 10% of sales or greater
- Capital Improvements: 7%–15% of sales
- Animal Feed: 10%–40% (depending on which animals you raise)

If you look at your bottom line and realize you're not making as much profit as you'd like (or need), then you can look at your benchmarks to see which areas of your business need improvement. Is your labor less than 35% of sales? Did you spend less than 3% of revenue on seeds?

To create benchmarks, you can start by looking at the scorecard developed by K. Becker, D. Kauppila, G. Rogers, R. Parsons, D. Nordquist and R. Craven. It is part of the financial workbook that can be found here: vhcb .org/Farm-Forest-Viability/resources/. After a few years in business, when you are operating successfully, you can create your own. You will know what the right volume of sales are, what is appropriate to spend on payroll, operating expenses and capital improvement. These benchmarks will also come in handy if you find your business slipping into the weeds: you will have a reference as to what it takes to run a successful business.

Measuring Success: Keep Your Eye on the Prize

You've put all this effort and energy into your business, and sometimes it can feel like you're spinning on a hamster wheel. How will you know when you arrive? When do you get there? And where is "there"?

When you first launched your business, you set out goals for yourself. When you feel mired down by the business or get in a rut, it's good to revisit those goals and see how you're measuring against them. This offers an opportunity to celebrate accomplishments, as well as refine your goals.

Having met initial goals in your business plan, you may want to establish new goals: higher sales targets, new sales channels, more efficient production and more vacations. Review your goals and set a target date for when you can continue measuring success.

Notes

1. In some regions, this may be considered a crazy high salary, and in others it may be below minimum wage. I fully support living wages, and at the time of this writing, it is considered a reasonable wage.

2. These numbers are based on an actual organic farm business. They may not be accurate for your specific business, location and climate. There are so many variables—from the width of your harvesting rows and walkways to the method of staking tomatoes to the type of potting mix and fertilizer you use. You will need to complete this kind of analysis using your own researched numbers that factor in your growing practices.

3. Richard Wiswall's *Organic Farmer's Business Handbook* offers a similar albeit more detailed method for calculating production costs. If you have followed his method in the past, and like it, great!

4. While most production costs can be isolated in QuickBooks, labor is more difficult, unless you have specific staff devoted to production (as opposed to selling, delivering, or admin). You may need to estimate the percentage of labor allocated to production.

5. To reiterate, these numbers are for illustrative purposes only and based on a single farm's experience. You would need to research and verify the numbers for your own business.

6. You can also calculate the cost per market. If you attend 3 markets per week, for a 20-week season, then the cost of the tents, tables and displays is $25/market (3 markets × 20 weeks × 2 years = 120 markets. $1,500 spread over 60 markets is $12.50 per market).

7. This breakdown of expenses is different than the earlier suggestion in the QuickBooks section. For day-to-day reporting and management, the earlier suggestion is better. For once-a-year analysis, this breakdown is better and easily adjusted from the QB reports.

8. Depending on if and how you round the numbers, you might get slightly different numbers here than I did.

9. Restocks only happen in the farm store.

10. For a refresher on how to enter the sales into QuickBooks, see page 113.

11. Right now, we're talking about raising prices on a few items, not your entire business. If, after doing this cost analysis, you realize you're losing money on all products *and* you can't raise your prices, then you'll need to reevaluate your whole business model. Where can you create value for your customers so you can raise your prices, and how can you cut expenses?

CHAPTER 9

Stabilizing Your Business

Sigi's Farm and Market

Sigi came across an incredible opportunity. She found a 100-acre farm for sale just outside Madison, WI, for $350,000. It was perfect! It had an old barn, a greenhouse, a farmhouse where she and her family could live and a farm store. The previous owners, who had farmed there for 40 years, had retired five years previous. They lived on the land hoping their kids would take over. The owners passed away, and the children decided to sell.

Sigi sat down and started crunching numbers: she had a pretty good sense of her customers and her ability to grow since she had already been farming on three acres. She thought about what she could grow, how many new customers she would need and how much she could sell in the farm store.

In her previous career, Sigi worked for a large food-service company, earned a good salary and saved close to $250,000. It had been her retirement savings, but she felt confident in her ability to run the expanded business; she was willing to risk the money.

Sigi hired a consultant to help her map out the numbers. They projected revenue and expenses for the different sales channels: wholesale, farm store and CSA. They laid out the start-up costs: what would it take to refurbish the barn and farm store? What new equipment would need to be purchased? How many months of operating expenses, payroll and mortgage payments did she need to make before she could expect to start earning revenue? She projected that the start-up costs would be about $125,000; she would put $50,000 down for the purchase of the land and take a mortgage for the remainder. She figured that even if start-up costs were more than she projected, she still had $75,000 left in her retirement account as a buffer.

In mid-November 2013, she closed on the property and closed her books. She didn't have time to look back at her projections: she had restoration to do! She started in the barn—securing the shelving and storage area. She continued in the farm store: cleaning, painting, retiling the floor; cleaning out the refrigerators and starting them up. She brought in a technician to ensure they were running properly. It turns out they weren't; after so many years idle, they sputtered. Sigi wrote him a check for $1,000, and he replaced the compressor. An unexpected expense, to be sure, but she knew she had wiggle room in her savings.

Sigi continued to write the checks—to buy new plastic for the greenhouse, fences to keep out the deer, a rototiller. Her business manager, Beth, who came with her from the previous farm, cautioned her that she was over her start-up projections and spending too much money. But what could Sigi do? If she wanted to get the farm up and running, she needed to make these purchases.

Finally, in May, Sigi and her crew culled the first few cases of radishes and arugula out of the ground. She hung the open sign at her farm store, called a few wholesale customers and started making deliveries. Sales were slower than she anticipated, but she was confident she could make it work. She continued to plant, harvest and sell.

Beth kept writing checks and asking Sigi for more money. The farm store shelves needed stocking to keep them looking full and inviting. New chicks were purchased to replace the ones lost to a predator. Sigi pulled money from her retirement account. In September, her accountant called: she had $5,000 left in her retirement account, and her checking account was bone-dry. With the next mortgage payment and payroll, her cash would be completely wiped out. Sigi realized she was in trouble and would need to get additional financing to get over this start-up hurdle.

When we looked at her books in November 2014, we discovered that she had not only run out of cash, but she also owed her vendors over $12,000! A traditional bank wouldn't give her a loan, so she needed to manage this cash crisis on her own.

Sigi was in the weeds, dug deep in a hole!

Managing the Hole

We set out to create a plan to get her out of the hole. While the long-term goal is to understand how to manage your business to never get in the hole again, there is no future if you can't manage the current situation and get out of cash-crisis mode. Sigi got there because she wasn't paying attention... and suddenly (though it didn't really happen suddenly) she found her bank accounts empty and credit cards maxed out. Vendors were banging at the door demanding to be paid. It's a scary place to be, but it is possible to get out of this hole.

The short-term NOW goal is to manage the hole and get out. So what can you do to get yourself out of the hole, and stay out?

My friend and colleague Denise Chew was the first person I heard articulate "The Rules of the Hole." She offered three rules, and I added a fourth. These are the rules by which you live if you find yourself in the hole.

- **Quit digging.** In other words, don't incur unnecessary or excessive expenses. Manage cash as if it were oxygen for a patient on life support.

  Identify your biggest expenses and manage them tightly. For most farmers, it's labor and direct operating expenses. Look at your labor schedule—do you really need all the staff working as many hours as they are? Are some employees racking up overtime? Look carefully at where you're spending money and make sure every dollar is carefully utilized.

  Figure out what systems you need to put in place in order to be sure these expenses are managed tightly. Create a budget for each day, week and month to decide what you can afford to purchase.

- **Keep the dogs at bay.** Manage and communicate with vendors (the dogs) about payment plans to keep your supply chain open. Often, explaining the situation to your vendors can be enough to buy you some time. If they know you have a plan in place, they will be more likely to work with you. Put together a list of payables and clarify the priority in which they need to be paid. You should prioritize according to the following:

- Ensure that you can continue basic operations. If you can't buy seeds, then you can't grow vegetables to sell. If you can't buy meat and dairy from the neighboring farm, you'll have no products to sell in your farm store.
- Minimize interest/late payment expenses. Credit card companies have more strict payment policies, whereas other vendors won't. Managing late payment fees will also help you to quit digging (see rule one).
- Understand and leverage your relationships with your vendors: are some vendors more flexible than others? Do you have good relationships that you can utilize to help stretch your payment terms for the time being?

- **Climb out.** The only way to climb out of the hole is to increase the money coming in, i.e., revenue. Focus on growing sales and actively sell, sell, sell. Solicit ideas from your team members (see Creative Thinking, page 175). Everything you do to cut costs and manage vendors will be moot if you can't drum up sales. What systems do you need to put in place to help team members (and you, the business owner) sell?
- **Get your head out of the sand.** Too often, when entrepreneurs get deep in the hole, they panic and start to ignore the realities of their situation. It is indeed possible to manage the hole and get out, but you must accept what's going on and be proactive. When you finally do get out of the hole, don't go back to your old ways. Continue to use the systems and controls that you have put in place so that it's less likely that you'll fall into this situation again.

There wasn't one mistake that got Sigi in the hole. It was a series of little things. Certainly, she did many things right—like creating financial projections to see if the plan was viable. But she forgot to look back at her projections and make adjustments as things changed; and she didn't look at her bank balance.

It was many little things that got Sigi in the hole, and it was many little things that helped her climb out. She was able to get a debt consolidation loan from her local farm service agency. This allowed her to pay off her vendors. She sold unnecessary equipment, streamlined her operations and cut back on payroll. In the farm-store, Sigi reorganized the displays to look full and inviting, even though she was selling through her inventory and storage crops. This allowed her to benefit from sales revenue with minimal production labor and without any additional purchasing. After 6 months of hustling sales and running a lean operation, she was able to build up a cash buffer and begin paying back her debt consolidation loan.

There are many little things that can be done, which we've discussed throughout the book.

1. **Have an effective bookkeeping system so you understand how your money is coming and going.** In order to manage cash flow, you must have an effective way to log each and every transaction, and classify them according to a system that makes sense to you. This allows you to extract the information you need. A detailed paper trail when the transactions occur is the only way you will realistically be able to aggregate and analyze the data effectively. QuickBooks is the most efficient tool to do this.

2. **Understand the root of your cash flow issues before it becomes a crisis.** If you ever start feeling like you're on the verge of a cash crunch, do some digging into your financial records to figure it out before the shit really hits the fan, and you can't, for example, make payroll. Here are a few common reasons why your cash flow may be suffering:
   - Customers don't pay on time. (see #7)
   - Poor planning for seasonally slow times. (see #5)
   - You have more debt than the business can sustain. (see #3 and #10)
   - You've let your expenses creep up (see #4, #6 and #9)

3. **Don't borrow against payroll or sales tax.** Every two weeks, you deduct money from your employees' paychecks to cover their contributions of payroll taxes (and you accrue a liability of the portion you, as the employer, owe); and every day, some businesses collect sales tax from their customers. These monies can hang out in your checking account for several weeks before being remitted to the government. Just because the money is in your account, doesn't mean it's yours! It belongs to the government, and you're just the messenger. If you're spending this money, you are essentially financing your business through a government tax loan. To avoid this situation:

- Have clear Taxes Payable accounts (meals tax, payroll tax) set up in your bookkeeping system and use them each time you log daily sales and payroll. When you look at your checking account balance, make a note to deduct the amount of payables.
- Better still, open a separate bank account to set aside the pass-through taxes. If you don't see it in your primary checking account then there's less temptation to spend it.

4. **Understand your profit model and what it takes to break even.** I've been working with a client whose business is not yet profitable, and she has borrowed money from friends and family to stay afloat. The challenge is she doesn't know where she needs to make changes because she doesn't understand the profit model of her business. That is, what does she need to do to become profitable? Are her expenses too high or revenues too low? While the answer to both questions is likely "yes," there are defined, targeted goals that she needs to achieve. It's not just: "increase sales," it's "increase sales to $XXX." It's not just, "reduce expenses," it's "reduce expenses to XX% of revenue." What are the target sales you need to cover your fixed costs like rent and insurance? What is the maximum you can spend on your variable costs like seeds and fertilizer? In order to stay profitable, and maintain good cash flow, you need to understand the benchmarks of revenue and expenses for your business.

5. **Know when your slow periods are and have a cash plan.** All businesses face a slow period during certain times of the year. Many crops stop producing with the autumn's first frost, and farmers will see a dip in cash flow from November until March when CSA subscribers start sending in checks. This can be a scary time if you haven't planned for the period of low cash inflow; after all, the rent (and other expenses) still needs to be paid. When you have a period of good sales and cash, be sure to squirrel away some for the low periods. If you have a cash flow plan, then you know how much extra cash you'll need to cover the slow times.

6. **Have a budget, and compare projections to actuals.** A budget for how much and when you expect to spend money and bring in revenues is a great tool to make sure you stay cash positive. But projections can feel like "fairy dust" if you only look back at them once a year to see how close (or how far off) you were. Throughout the year, review your projections and make adjustments. If you see that expenses are higher than antici-

pated, then you'll need to cut back in the coming months to stay on track. If your revenue is better than anticipated, then perhaps you can make the capital improvements that you had otherwise delayed. Further, adjusting your budget for the coming year according to your historical revenues and expenses will help you home in on the realities of your business's cash flow and can help you avoid a cash crisis.

7. **Manage your Accounts Receivable.** Accounts Receivable (or A/R, money that your customers owe you) can turn into a cash crisis quickly if not managed effectively. They are easy to forget about when managing day-to-day operations; you know you have made the sale, and you have delivered the product. But if you don't have a clear policy in place on when payment of invoices are due or if you don't invoice your clients on time, the actual payments can come trickling in more slowly than you expect, if at all. If you have customers that consistently don't pay their invoices on time, it's completely reasonable to stop doing business with them. After all, it could mean the difference between you being cash solvent or in the hole.

8. **Always look for ways to reduce expenses and increase revenue.** Perhaps this one is an obvious Business 101 lesson, but it's not always easy to think outside of the box when you are in a rhythm with respect to your business operations. Having trusted advisors that can help you identify potential areas of expense reduction and revenue enhancement can be a valuable way to keep your cash flow in a healthy positive state at all times.

9. **Understand the hidden costs in your products and price them accordingly.** You've gone through your books, and you calculated the costs to grow and sell a case of arugula. Beyond that, entrepreneurs often miss a few other areas when thinking about prices:
   - How much do you waste? A sad reality is that not everything you grow or make will get sold, and much of it will end up in the compost bin. Factor in your "shrink" when setting your prices.
   - Are you paying a sales commission? While this can drum up sales, it also eats into profits.
   - Have you factored in the cost of packaging and shipping?

10. **Never borrow money if you can't afford the monthly payments.** Borrowing money isn't a bad thing. Often it's the most efficient way to grow your business. But before you borrow money, make sure you can afford

to pay it back (both the principle and interest) through the profits of the growing business.

This is especially true of credit cards. They can be a convenient way to pay for goods when you don't have cash on hand (and the points you earn are a bonus!). But if you can't afford to pay them off at the end of each month, then you will rack up serious late payment and interest charges.

When to Stop

Throughout this book, we've discussed many measurements for success; and as an entrepreneur, you probably thought of many other ways—reaching a revenue goal, customer base or profit. One measure that we have not yet discussed is making the hard, though sometimes right, decision to close up shop. As entrepreneurs we are fighters, we scratch and claw to be successful, we do things differently. And if we don't attain the traditional measures of success, does that mean we're a failure? Not at all.

Success is recognizing your economic reality and making the difficult decisions. And if there are lessons to be learned, seeing them, getting back up and applying the lessons to your next venture.

The decision to close up shop may come from having dug yourself into a deep hole, or from the realization that you're not climbing towards success as fast and as high as you'd like.

Blue Moon Farm

Deirdre from Blue Moon Farm grew organic produce and grains and raised chickens. The eggs and grains went into her homemade pastas, and the vegetables were used for ravioli fillings and sauces. During the summer, she and her crew frantically managed the farm and the kitchen to make the most of the growing season. Her products were more expensive than those found in the grocery store, but she felt confident about the quality and taste of her pastas. She sold her pastas and sauces in her farm store, which was in a lovely residential neighborhood. After six years in business, Deirdre still wasn't earning a decent living, and continued to rely on her husband's salary to cover the household expenses. She called me to help her figure out how to make it work.

Like most entrepreneurs, she had been so focused on the day-to-day operations that she hadn't done much with her bookkeeping. Sure, she paid her bills and filed her taxes; but beyond that, she didn't have much to offer me in terms of her historical financials. In fact, her accountant wasn't that good either, and had missed filing the meals tax. As a result, Deirdre had just finished recovering from paying back taxes.

In order for us to figure out how to make her business thrive (and not just survive), we needed to get her bookkeeping in order. With a few months, or more, of solid records, we could understand the economics of her business and determine what changes needed to be made.

A year later, Deirdre had excellent financial records to find the trouble spots. We reviewed her revenues and expenses and discussed what changes needed to happen to create a thriving business. We identified months that were profitable and months that were not. In comparing the profitable months with the non-profitable months, we set targets for revenue, payroll and direct operating expenses. We brainstormed ideas for increasing sales and discussed strategies for tempering her costs. We agreed that we would revisit her financials in three months to see what kind of progress she had made towards those goals. Then, she could make a go/no-go decision.

After three months, she had barely moved towards her goals. She was exhausted and emotionally drained. She decided it was time to close up shop.

If you're in the hole, you must first ask yourself: is it, in fact, possible to get out? If you're not making enough money, is it possible to grow? Is it worth the effort to try? While being "too deep in the hole" or "not successful enough" is certainly relative to each entrepreneur's resources, there are five critical questions to ask when considering whether it might be time to close up shop:

1. **Do you have what it takes to be a successful entrepreneur?** We talked about many of these in Chapter 3: Planning Your New Venture. At the

core of those seven items: you must love what you do and take it seriously. But successful entrepreneurs succeed in the long run by having the ability to manage money wisely, ask for the sale, build a top-notch team, be accessible, negotiate successfully and stay organized. After a few years in business, you may realize that you're not cut from entrepreneurial cloth.

There's also a big difference between what it takes to be a successful entrepreneur in start-up mode versus in business growth mode. There comes a point where you need to start delegating. All the technical skills that got you started successfully (whether it's how to work the soil or build a customer base) won't help you get to the next phase of growth. Here you need to be able to transfer your skills and vision to your team, so you can work together to grow the business. If your business is struggling, it may be that you need to bring in a CEO to help you take your business to the next level (see Chapter 10: Growing Your Business).

2. **When you pull debt out of the equation, do you have a sustainable business model?** If you've gotten yourself in debt, and it's getting deeper, it's important to understand why. Is the burden of debt digging you deeper, or is the business model at its core not profitable? In other words, if you ignore the debt payments, can you cover your costs solely from your operating revenues and without infusions of cash from loans or grants? It's critical to understand your operating income outside of non-business related revenue and expenses. You obviously can't rely on loans in perpetuity to keep your business afloat. Likewise, it's unsustainable as a for-profit business to rely on grants—you never know when charitable funding dollars might dry up.

Often, an entrepreneur's income statement has interest expense and debt repayment buried within the operating expenses. By pulling them out, you can see what the income statement would look like if you didn't have any debt. Without the burden of debt, do you have a profitable business model? (See Chapter 6: Setting up QuickBooks.) If the answer is yes, then there may be opportunities to manage it and dig out.

3. **If you look back at your historical financial data, can you see what success looks like?** It may very well be that right now you don't have a profitable business model. After all, something happened that led to

mounting debt. Beyond the existing debt burden, review your historical financials, either by month or by year, to see if you can identify a period when you were profitable. When you do see a time when business was profitable, think about what made that time different than today: Were you operating on a smaller scale or with fewer employees? Was your revenue higher? Were your expenses lower? Did you sell different products or sell through different channels? Can you replicate the circumstances that made the business profitable?

Templeton Florists

Dean was a third-generation flower grower; his family started the business 80 years ago just outside of Philadelphia, PA. Having grown up on the farm, Dean knew flowers. He knew how to create bouquets and manage a delivery schedule. He had wholesale customers and established relationships with funeral homes (a surprisingly large portion of the business). When his father retired, Dean naturally took over the family business. Truth be told, Dean didn't have much of a head for numbers, so he hired a bookkeeper to track the sales and pay the bills. He didn't really worry about it—after 80 years of business, the farm and flower shop pretty much ran themselves. It wasn't until about seven years after he took over the business that he realized things weren't going quite as well as he expected. His credit card debt was rising, he had a second mortgage on his house, and he hadn't paid himself in years. As he approached his own retirement, he knew that things needed to change. He needed a business that he could sell or that could support him in his retirement.

When we met to figure out a plan, the first thing I wanted to know was what success would look like. What would Dean's business look like without the burden of debt? He was paying close $30,000 a year in interest alone! We knew that the for the last year or so the business wasn't profitable, but could we find a time when it was? What did that look like? We looked at the last 10 years of the business; 2007 had been the best year. In that year, total sales topped $600,000, his cost of goods was 30%, and payroll was 43%. For Templeton Florist to succeed and pay down debt, he needed to better those numbers—targeting sales of $700,000 with 25% COGS and 40% payroll. With a firm target in hand, Dean created a strategy to meet those goals.

4. **Can you support your debt service?** If you are struggling to get out from under your business debt, the first thing to do is communicate with your creditors. See if there's an opportunity to work out a plan to consolidate the debt and not accrue additional penalties and fees. You can then create a debt service plan (see Debt Service Template at juliashanks.com /TheFarmersOfficeTemplates/). Once your payment plan is in place, be sure that, based on your historical revenues, you can reasonably project to earn enough operating profit each month to make at least the minimum monthly payments on your debt. Based on what you can afford to pay toward decreasing the debt each month, how long will it take to pay down all of the debt in full? Can you afford to wait that long for the business to once again generate substantial cash flow? If your operating income isn't going to be high enough to keep up with your debt service payments, it's time to reassess your business model as a whole.

5. **What is the consequence of waiting?** Think about the worst-case scenario if you continue to operate. What will you sacrifice? For example, do you need to tap into your retirement funds significantly, with retirement age is just a few years away? In Sigi's case, she had depleted her entire savings account. For Dean, his ability to sell his business would diminish if he wasn't able to recover sufficiently.

   Or perhaps you have a strong brand now that you can leverage to sell, but that you won't have in a year's time as the business continues going downhill and the brand degrades.

   When you're in the hole, it's important to be realistic with yourself about your business's situation and prospects for the future. Denial will only cost you more in the long run. (See rule #4 of managing the hole.)

Key Takeaways

The hole is a scary place to be and can turn into quicksand without a quick recovery. To avoid the hole in the first place:

- Review your business model to ensure it is profitable and can weather unexpected challenges.
- Create a cash flow plan and review it monthly (at a minimum) to ensure you're staying on track.
- The moment you realize you're off-track, take corrective actions to increase revenue and reduce expenses.
- Never borrow money if you cannot create a realistic plan to pay it back.

With a stable business, you can think about growth.

CHAPTER 10

Growing Your Business

Most farm entrepreneurs continually plot and strategize growth, whether to expand the number of acres in production, create value-added items or open a farm store. After all, the successful growth of their business means the potential to earn more profits—and who doesn't want that?

As you think about growth, the key to your success lies in your ability to make thoughtful decisions to ensure new projects stand on their own, while maintaining the stability of your existing business.

Establish Goals

When you initially launched your business you may or may not have set goals for yourself. Like many entrepreneurs (myself included), you may have just jumped right in, fearing that if you thought too much about it, you'd never start. Or maybe you really didn't know what kinds of goals to set. Here you are now, thinking about growing your business, and you want to be more strategic about it. And to do that, you must first lay out your goals. If you want, look back at your earlier business goals. No doubt, you have new ideas of where you want to go.

Why do you want to grow your business? You may want to improve your quality of life, and that means earning more money. Beyond earning more money, you may also want to spend more time with your family. That means your growth strategy should allow you to earn more money without necessarily working more hours. You may want to grow the business sufficiently so that you have something to pass on to your children. That means understanding where your kids' interests lie, and sufficiently establishing the business for when the kids take it over.

State three goals in growing your business:

1. _____

2. _____

3. _____

With clear goals, you have a reference point to ensure that the strategies you pursue align with your goals. Growing for growth's sake is unproductive, since you'll never know if you've succeeded.

Deciding How to Grow

You decided to grow your business. You've been successful so far—people love your products—and you want to make more money. There are so many ways to grow a business:

- Streamline and/or expand current operations
- Add more products
- Start a new enterprise.

However you grow your business, think about growing the top-line revenue as well as the bottom-line profits. In other words, figure out what makes you money.

Organic Growth

In business school, they taught us that successful businesses always grow. If you decide to stay small, this classifies you as a "lifestyle" venture—which has a derogatory tone to it. But there's nothing wrong with staying small: growing too big or too fast can lead to unsustainable practices and pressures that you may not want. Bigger is not always better; don't let outside advisors pressure you into making the wrong decision for *you*.

Stop

Growth isn't always about doing more. Just because you're increasing the top-line revenue doesn't mean you're seeing the proportional increase in net income.

When Bigger Isn't Better

For 20 years, Farmer Dan from Solstice Farm grew field and green-house lettuces that he sold to wholesale distributors. Twice a week, he sent his trucks two hours from his farm in Central Illinois to Chicago to make the deliveries. In the depths of winter, he supplemented his greens with product from California. This kept his customers well stocked until Dan could ramp up his own production in March. While this was his method for many, many years, he finally sat down to do the math: with the cost of the delivery run plus the lettuces he purchased for resale, he actually lost money. The next year, he decided to sell only his own product. The farm went from $150,000 in winter sales to $35,000; but instead of losing $10,000, he earned profits of $10,000.

If you can figure out what to stop doing, you'll be able to redirect that energy, time and effort into things that work. If a product, service, location or partnership is dragging the rest of your operations down, maybe it's time to think about moving on, redirecting and prioritizing your opportunities.

Fail Better

What do you regret from the last year or so of business? What were your most embarrassing failures? Once you stop cringing, think about what you'd do differently next time. A failure can be as simple as over-ordering seeds or planning your delivery runs during peak traffic.

Assess your failures and setbacks, learn from them, and fail better next time.

Expand Operations

Often you'll have a gut instinct as to what your most profitable items are, but as one client said, "I don't have enough confidence in my gut to make decisions. I need to see the numbers on paper."

In Chapter 8 (Digging into the Numbers and Beyond), we talked about cost accounting—how to understand the true cost of your products. From an operational perspective, you want to make sure that you are making money on all your products and pricing them appropriately. From a growth perspective, you want to know what makes the most money so you can decide how to grow.

Which Products?

When Susan and Olivia at Wingate Farm (from the Introduction) decided to grow their business, they explored both meat birds and egg production as potential opportunities.

As they evaluated the meat-bird production, they calculated the time they spent managing the birds (daily feeding and maintenance, plus slaughtering) and the direct expenses. We looked at the year as a whole:

- How much time did they spend on the daily chores of watering and feeding the birds?
- How many hours did they spend slaughtering chickens?
- How many chickens did they slaughter at one time, and how many times in a year did they slaughter chickens?
- How much money did they spend on feed, chicks, packaging and other supplies?

They added up all the expenses, along with the cost of the labor; and divided that by the number of birds they sold. This gave them the total cost to "produce" a meat bird. With a sales price of $20 per bird, they were making just a few pennies per bird. This informed their growth strategy—if they wanted to keep selling chickens, they needed to raise their prices and find a way to reduce their expenses.

On the other hand, the eggs were quite profitable, even when they sold them for $5.50 per dozen.

We added up the number of hours they spent on their chores: feeding and maintaining the birds, gathering eggs, washing and packaging. At 732 total hours for the season, that was $10,980 in labor costs (732 × $15/hour). We looked at their QuickBooks files and added up all their expenses associated with the birds; it was $14,620. The total cost of producing 5,880 dozen eggs was $25,600; or $4.35 per dozen. See Table 10.1 for a breakdown of the numbers.

For Susan and Olivia, it made more sense to expand the egg production. Their growth strategy focused around finding new channels (and customers) to sell eggs.

TABLE 10.1. Cost of Egg Production

Expenses	Season Total	Cost	Unit
Labor Hours			
Chores	372	$0.95	dozen
Washing & packaging	336	$0.86	dozen
Misc.	24	$0.06	dozen
Direct Expenses			
Pullets	$1,980	$0.34	dozen
Packaging	$2,047	$0.35	dozen
Feed	$9,794	$1.67	dozen
Winter supplies	$208	$0.04	dozen
Coop repair	$66	$0.01	dozen
Supplements	$133	$0.02	dozen
Other	$392	$0.07	dozen
Total Costs	$25,600	$4.35	dozen

Which Sales Channels?

Susan and Olivia's decision to increase the egg production was easy. From there, they needed to figure out how they were going to sell all the eggs. Do they attend another farmers market, find more wholesale customers or increase the size of the CSA? Like the analysis we did in to decide *what* to grow, you can calculate the cost to sell through each channel.

Sales costs include:

- **Labor:** time spent talking on the phone with your customers and emailing them, in addition to delivery time
- **Delivery costs:** How many miles do you travel? How much do you spend on gas and tolls?
- **Other costs:** Is there special packaging for certain sales channels?

In addition, to enter a new market, there are start-up costs. It could be brochures for prospective CSA customers, sell-sheets for retailers, or tents and tables for a new farmers market. These costs can be spread out through the course of a season or a few years.

To sell at the farmers market, the costs are:

- **Labor:** Your time plus the time of any employees who will be working with you. Assign a wage to yourself, as well as the employee. After all, if you're not doing the work (and you may not as your business grows), you will pay *someone* to do it for you. This includes time to pack the truck, drive to the market unload the truck, set up the stand, sell at the market, break down the stand, load the truck, drive back to the farm and unpack the truck.

- **Packaging:** All those cardboard pint and quart containers that you display your produce in, plastic shopping bags that the customers take away with them.
- **Mileage costs:** Fuel to drive to and from the market, plus wear and tear on the vehicle.
- **Market entry fees:** What does it cost to start selling in this channel? If selling at a farmers market is new to you, then you need to invest in a tent, table and displays. If you're selling to wholesale customers, you will spend many hours calling restaurants, markets and distributors who would buy your product.

For Susan and Olivia, selling to wholesale customers cost less money, but they earned less per dozen. It made more sense for them to grow the CSA sales.

Consider Financing

Maybe you haven't yet had to borrow money, relying solely on savings and friends and family. However, to grow to the next level, even if it's a small step, outside financing may now be required. Financing can help you develop your current offerings, expand to new ones or streamline your current systems.

For Susan and Olivia, the $7,000 investment in an egg-washer, nest boxes and other equipment for their egg production was worthwhile. The equipment would last three to five years, and would allow them to increase sales by 60%. In addition, the egg-washer would reduce their labor cost as a percentage of the total cost of producing the eggs.

Ask yourself: "Would the purchase of new equipment allow me to increase revenue or decrease expenses? Is it worth making that investment?"

(We'll get to that analysis in a few pages.)

Add New Products and Services

Listen to your customers, and I mean *really* listen; they will tell you what they want. Maybe your customers want a more customizable CSA, different cuts of meat than what you currently offer or value-added products.

To get to the next level of growth, it might take something different and unexpected. It can be intimidating to depart from what you know to work so far, but taking the plunge into new possibilities might be just what the doctor ordered.

For example, if your farm has only 25 acres, your revenue is capped by how much you can cultivate. You'd need to figure out a way to increase revenue beyond the basic operations in order to realize your goals. Maybe that's growing higher-value crops, creating value-added products or offering "agri-tainment."

Some of the customer needs you can meet on your own—maybe you contract grow for restaurant customers. Other times, increasing your offerings means partnering with another farm. As a vegetable farmer, you may want to partner with a meat farmer and grain farmer to offer a full-diet CSA. Another example could be to partner with a chef to create value-added products.

Start a New Enterprise

Maybe you have a gut feeling about what you want to do next—perhaps it's creating a line of value-added products with your produce or building a farm store. If you don't have an idea, but have an itch to do more, brainstorm.

─────────── Creative Thinking ───────────

Sometimes business challenges arise that require more than just digging into the numbers. Perhaps it's an operational challenge or trying to figure out how to grow your business. Brainstorming can be a good way to rally your team around new and creative solutions.

Brainstorming is most effective when you foster an open environment that allows participants to stretch themselves. It's often the crazy stretchy ideas that spark the more realistic but new and exciting ideas. Here are two ways that you can improve brainstorming sessions:[1]

No Bazookas

Imagine you're in a brainstorming session with your crew and you offer a suggestion. A co-worker chimes in: "Well that's the dumbest idea I've ever heard!"

I'm sure I don't need to tell you that shooting down someone's idea is not a productive way to interact with colleagues or clients. Two things happen as a result. The first, and probably more obvious, is that the person who had the "bad" idea is less likely to contribute more ideas. I won't go so far as to say, "No idea is a bad idea"— there *are* bad ideas—but that person could very well have other "good" ideas later on. If he fears his idea will be bazooka'ed, he's less inclined to contribute. The second, and subtler, consequence is that all ideas (good, and especially the bad and absurd) can spark further ideas—driving a string of thoughts that could lead to the great idea.

Let me give you an example. In a brainstorming session with a client, I offered an unpractical idea: *I wish you had wine breaks instead of water breaks.*

This silly idea generated other ideas around creating a social environment with friends, which in turn sparked the notion of date nights at the farm. The idea started growing legs and led to creating a social event around the CSA pickups—to share a glass of wine and recipes. Customers would linger on the farm and were more likely to purchase items from the farm-store.

Open-minded Evaluation

When going through the process of brainstorming, you will come up with many ideas, several of which you will want to pursue further. As you evaluate ideas to pursue, it's important to recognize that no idea is perfect. Rather than toss out every idea that isn't perfect, or immediately feasible, take the opportunity to evaluate the idea with an open mind towards making it possible. The trick is to identify the flawed but exciting "beginning" ideas and have a thinking tool for highlighting the parts you like and overcoming the negatives.

Like most ideas, the CSA social pickup wasn't 100% perfect or 100% dud. Its potential for perfection increased as its feasibility increased. The trick is to take a new idea and push it along the spectrum of perfection by getting it past the threshold of feasibility.

It's easy to come up with all the reasons why an idea won't work, but that will not make the idea more feasible. Instead, give it some energy by starting with all the ways in which the idea is a good one. For example:

- It builds community among our members, and loyalty to our farm.
- It increases sales at the farm store.
- It increases the probability that our customers will tell their friends about our farm.

By starting with the positives, the idea generates some energy to push it through the issues that it may have.

And instead of just creating a list of why an idea won't work, invite some problem-solving. One easy technique is by stating your concern, starting with the two most powerful words in problem-solving, "How to." For example, one issue is: *These socials will cost too much money.* This issue can be restated by starting the question as a "How to": *How to get the socials to pay for themselves?*

By the simple restatement of the issue, it encourages us to think about the issue in ways that it can be resolved, and this will give the idea enough momentum to make it feasible.

Whatever you decide to do, a new enterprise warrants revisiting the business planning process. As you know, this helps you flesh out all the details that go into the new enterprise.

Reflect

After all the talking with your customers, brainstorming with your crew and analyzing your costs, you've come up with an idea. Great! Look back at the goals you laid out in the beginning of this chapter. Does your growth idea align with your personal and business goals? If so, you can continue with the planning process. If not, consider how you can adjust your idea to align with your goals.

Clarify Your Vision and Create a Plan

You decided on the best strategy, and it aligns with your goals and values. Maybe it's increasing your acreage in production, building a greenhouse and creating a value-added product. So how do you get there?

Establish Concrete Goals

Growth strategies become much more productive with concrete directions and a road map. Say you want to expand your vegetable production; by how many acres? How many new customers do you need to make that happen? How many new markets do you need to sell at? Once you establish the concrete end-points, project what you need to accomplish them, and by when.

Make Success Possible: Create Milestones and a Timeline

Getting to an end point can be daunting without more manageable milestones. Chunk it down: set goals you can achieve, and celebrate each one when you get there. Here are a few examples:

- **Increasing farmers market sales:** In order to grow sales by 50%, you might need to apply to more farmers markets or create a crop plan to increase production. You may want to grow 25% for two years in a row. Set smaller, achievable milestones within your time frame to clarify how realistic your plans are, and help you realize what you need to do to get there.
- **Increasing CSA revenue:** In order to increase your customer base, you can set seasonal growth goals of 25 new customers per CSA type. You can also list strategies to attract more customers.
- **Creating a value-added product:** You may need to first expand your vegetable production, then find a commercial kitchen to produce your product. You may need to run a few test batches before you can scale up in production.
- **An agritourism plan** may call for partitioning part of your fields in stages, creating parking and walking areas, and creating marketing materials.

What Do You Need to Make It Happen

What investments do you need to make to realize your growth plans? For Susan and Olivia, they needed to build winter housing for their birds. For agri-tourism, you may need to prepare new beds for production, work with your town and state officials about highway signage and hire someone to create a parking area.

Take your list of needed investments and write down the cost associated with each item.

Running the Numbers—Enterprise Analysis

You know how you want to grow your business and where you will sell your goods. You know what kind of investments it will take. You can take a break to run the numbers. Sure, the greenhouse will increase production, but is it enough to justify the expense? The egg-washer will increase efficiencies, but are the labor savings enough to justify the cost of the purchase?

If you're already running a profitable business, then you don't necessarily need to write a full business plan and create full financial projections. You just need to look at the area where you want to grow, understand its impact on your current operation and create an enterprise analysis.

With a new enterprise comes new expenses (and of course the intended revenue). Some of these expenses are variable and some are fixed. They may overlap with some of the expenses you already have. If you decide to erect a new greenhouse, you may need some sort of fuel to heat it in the winter and soil amendments to replenish nutrients in the spring. Fuel and amendments may very well be a part of your existing operation. As you explore the viability of a new enterprise, outline the additional expenses associated with it.

Wingate Farm decided they needed an additional egg-washer, nest boxes and a chicken coop in order to grow. They knew that the eggs were profitable if they already had the infrastructure, but they weren't sure if they could afford the investment. They ran the numbers:

1. List out the start-up expenses:
   - Egg-washer—$2,000
   - Nest boxes—$3,000
   - Chicken coop—$2,000
2. What would it look like if they borrowed money to finance the purchases:
   - A 3-year loan at 6% interest translates to monthly payments of $244.
3. With the additional purchases, they figure they could yield an additional 4,000 dozen eggs per year. At a price of $6.25 per dozen, that's
   - $6.25/dozen * 4,000 dozen = $25,000
4. Estimate new expenses:
   - With the new egg-washer, there will be additional operating costs—water, electricity and sanitizer. They figure that will cost about $240 per year, or $.06/dozen (1% of the sale price).
5. Consider how your expenses will change with increased production. You can look at the expenses as a fixed cost, or as a percentage of the sales price.

Susan and Olivia looked down the list of expenses and considered how they might change with the increased production. For sure, they felt their labor would be more efficient, especially around egg washing. They also called their vendors to see if they could get a reduced price on their feed and packaging since they would be purchasing a greater volume.

With these efficiencies, they figured their variable cost with delivery would be 68.5%, down from 77%. If they sold $25,000 worth of eggs, the variable costs would total $17,115. That leaves them with $7,885 in contribution margin, that is, the additional profit earned for the business as a result of the increased production.

		Historical Cost as Percentage of Revenue	With Increased Production
Labor Hours	Chores	15.2%	14.0%
	Washing and packaging	13.7%	9.0%
	Misc.	1.0%	1.0%
Direct Expenses	Pullets	5.4%	5.2%
	Packaging	5.6%	5.0%
	Feed	26.6%	25.0%
	Winter supplies	0.6%	0.6%
	Coop repair	0.2%	0.2%
	Supplements	0.4%	0.4%
	Other	1.1%	1.1%
	New Expenses		1.0%
	Expenses Before Delivery	69.7%	62.5%
Delivery	CSA (Labor)	6.1%	5.0%
	Fuel	1.3%	1.0%
Total		77.0%	68.5%

They compared the contribution margin to the cost of financing. The monthly payment of $244 totals $2,928 per year. The contribution margin ($7,885) less the cost of the debt would provide an additional $4,957 annually to the business resulting from this opportunity.

	Numbers	Calculation behind Numbers
Revenue	$25,000	$6.25/dozen × 4,000 dozen
Less Expenses	–$17,115	68.5% × $25,000
Contribution	$7,885	$25,000 – $17,117
Less Debt Service	–$2,928	$244 × 12
Contribution to Business	$4,957	$7,885 – $2,928

Certainly there are other expenses in their business—rent, utilities, liability insurance and so on. But these are sunk costs; whether or not they expand production, they will still have these costs. Therefore, they should not weigh into determining the viability of the new enterprise.

Wingate Farm's growth strategy did not require a lot of investment, and an enterprise budget along with a Quick and Dirty Cash Flow Projection (juliashanks.com/TheFarmersOfficeTemplates/) is sufficient. For larger expansions and growth strategies, you'll want to create a full business plan and financial projections (see Chapter 4: The Business Planning Process).

Capitalize Appropriately

Remember Michael from Six Hands Farm (Chapter 2: The Financial Statements)? He made the mistake of borrowing more money than he could afford to repay. It's one thing for a business to be able to operate profitably at its core: the business generates more revenue than it incurs expenses (as he did). But if you need funding, then the business must generate enough revenue to cover expenses *and* the cost of the financing. Even if the banks loan you more than you need, make sure you need it and can afford it.

On the other side, the entrepreneur needs to borrow enough money to get through the start-up phase. While you don't want to borrow *too* much money; you need to make sure you borrow *enough*.

On the Move Mobile Market

In February 2013, Jackie and Steve began retrofitting a school bus into a mobile farm store. They had borrowed $50,000 to purchase the bus and install a custom air conditioning unit to ensure the produce was in a temperature-controlled climate. They were scheduled to begin selling in April 2013. Unfortunately, midway through the process, they ran out of money. The shelving units they planned to purchase didn't work, and they needed to buy more expensive units. This would delay the project several months while they raised more funds. Further compounding their troubles, Steve ran out of personal savings and needed to get a job (he had already quit his day job in anticipation of the new business launch). As close as they were to launching, they had to abandon their project for a year, while they recapitalized themselves and the business.

As you plan your venture, you want to make sure you have enough money (capital), but not too much, to effectively finance and launch your business. As part of the planning process, I recommend that you budget extra money to accommodate unexpected delays in launching and unexpected expenses. "Expect the unexpected." It will happen, so you might as plan for it.

If you're creating financial projections to test the feasibility of your plan (and not just trusting your gut), then you can home in on the right amount of funding needed.

After careful consideration, you decided how you want to grow your business. You may opt for self-funded growth or outside funding. All the back of the envelope number crunching you did above is enough to make a decision, but if you require outside funding, then you'll need to write a business plan and create financial projections. The scope of your expansion will dictate the size and scope of your business plan and financials. (See Chapter 4: The Business Planning Process.)

Develop New Skills

The skills that brought you to this first level of success are not necessarily the same skills that will bring you to the next. As a start-up entrepreneur/farmer, you bootstrapped, scrimped and saved. You were in the fields weeding and harvesting, at the farmers markets selling and delivering to your wholesale customers.

As you grow your business, you will need to hire people to help you. With just 24 hours in the day, you'll need to start delegating…and hone your HR skills—learn how to hire employees that are a good fit for your farm as well as having the technical skills; train; delegate; and sometimes fire. As CEO of your farm and business, you'll devote less time on the day-to-day and more on big-picture operation. You'll learn new skills in your new role as well as learn to delegate your old responsibilities.

Figure out where your skills are best utilized and expand your team to fill in the gaps. This can be in the form of partnerships or hiring staff.

- If you can identify specific needs, like marketing or graphic design, cultivate a partnership with another specialist. You might even be able to barter services for food.
- If you're already working at capacity, expanding your business could mean a workload greater than you can handle with your current staff. It

might be time for you to add some new faces to your business and delegate some of your chores. Start small: you don't need to delegate whole segments of your business at once, but if there are discrete tasks you can let go of, you'll be freer to focus on how to better grow your business.

Tangerini Farm

After 30 years farming, Laura and Charlie Tangerini recognized that they needed to shift their roles. They no longer had time to be production managers; they had to focus on general business management. As the general managers, they could create a strategy for the future: decide how to continue growing the business and develop the plan to execute upon it.

Laura knew one thing for sure: she didn't want to manage the farm crew anymore, nor rely on college and high-school students. She and Charlie decided to hire a farm manager: a person to lay out the crop and production plan, hire a crew and get things picked, washed and put away. The manager would scout the fields once or twice a week to see what needed to be done. After consulting with other large farms in the region, they settled on $60,000 as the appropriate salary for this new position.

Before hiring a farm manager, they had to figure out if they could afford this person: could they increase revenue sufficiently to justify this new expense? They reviewed their financials line by line to understand the opportunities for growth and ways to cut expenses. After some serious number crunching and business planning, they were confident that if they focused on growing the business, they could afford to hire for this position.

With their newfound time, they worked on branding materials, farmers market brochures and seasonal promotions. These were all the things they didn't have time to do when they were in full-on production mode; the things that would enable them to grow the business. After a year with the new production manager, Laura commented that they generated more profits than they had in any year previous. What a remarkable feeling to have more cash in the bank than they needed for the coming year!

Delegation

No one does it quite like you. As an entrepreneur, you take pride in how you do things…your unique skills created this successful, growing business. Ideally, you can clone yourself, so that all the people you hire do things exactly like you. Unfortunately, cloning technology is not here yet, so you need to learn to delegate and recognize that other people will be handling "your baby." They won't do it just like you.

The hardest part of growing a business is learning how to delegate effectively. Here are eight tips:

1. **Decide what you will delegate.** Throughout the course of your day, you do many, many things: from calling chefs to take their order to packaging CSA shares to selling at the farmers market.

   a. Make a list of all the tasks that you complete in a given week, including the tasks you *should* be doing as well as the ones you actually do. You can further organize this list into production tasks, sales and marketing tasks, and business management tasks.

   b. Note how much time you spend each week on each task. If it's a chore that you skip because you don't have time, note how much time it would take if you did have the time.

   c. Rate each task. Twice. First, on a scale of 1–10, where are your strengths? Which tasks do you rock (10) and which tasks do you barely get done, if at all (1). Second, rate the tasks in order of which absolutely must be done by you vs. it's okay if someone else does them—10 being most important, 1 being least.

   d. Go through the list of tasks, add up both scores for each task, and then put the list in order from high score to low score. The tasks at the top are the ones you totally rock and really need to be done by you. The tasks at the bottom are the thorns in your side that anyone but you should do.

   e. Look at the tasks in the middle—the ones that you are pretty good at, not great, and maybe need to be done by you. Do you need to reprioritize the list a bit?

   f. Now start at the top of the list and start adding up hours. When you get to 40 hours, stop and draw a line. Everything above the line, you do. Everything below the line, you delegate.

"But wait! Julia, c'mon…I work way more than 40 hours a week!" I know, right?! The remaining 20–40 hours a week will be managing your crew and keeping them on track, hiring and firing, and dealing with all the little emergencies that crop up during each day (crows ravaging the watermelon field, the tractor that just broke, etc.).

2. Once you've decided what you will delegate, clearly **define roles and responsibilities.** Is this a new job or new tasks for a current employee? Is this a role for one person or multiple people? Create new job descriptions:
   a. What tasks are expected—what exactly are you delegating?
   b. What are the skills required for the job?
   c. To whom will the person responsible report directly? To whom will they report indirectly?

3. **Communicate** to the person taking on the new role:
   a. What are your expectations for how the project/job will go?
   b. When do you expect tasks to be completed?
   c. At what point should questions be asked?
   d. What does success look like?

Side Note: When working remotely in the fields or on a delivery run, your employees will find it helpful if you let them know how you like to communicate. Do you prefer text messages or phone calls, or do they need to find you?

4. **Create feedback loops:** don't just send your employees off and expect things to go exactly as you planned. Check in on a regular basis to make sure things are proceeding as expected, answer questions, and make adjustments as necessary. (But try not to be overbearing!)

5. **Have clear expectations.** If you want a specific result, you have to ask for it! Always articulate the goal, along with any specific details that need to make it into the final product on the defined schedule. How ripe should the tomatoes be when you harvest, how big is the ideal okra pod? If you're not sure what the end product should look like (what's the best recipe for your homemade ketchup, for example), ask for a few different options to choose among, early in the process, and discuss which direction is best for the rest of the project. Make sure you confirm the task, goal and timeline with each other to prevent miscommunication down the line.

6. **Build in accountability.** Planning regular check-ins lets your employees know there's support available and that they can ask questions as needed. This also allows you to redirect the process as needed or as your farm

needs change. Adding check-in points also makes it easier to break a big independent project into smaller steps, with specific goals along the way to keep on track.

7. **Consider employees' growth.** Use new responsibilities as a way to develop your employees' own capabilities, not just as a way to lighten your own load. By developing their skills and strengths, you invest in making your staff more valued. Further, if your employees continue to learn and grow, they are more likely to stay in your employ. Delegation should serve as on the job training that benefits both of you.

8. Finally, **be open** to new ways of doing things: the task might not get done exactly the way you'd imagined it, but that diversity and new direction often proves to be a strength. You can do everything yourself and stay small, or you can grow on to bigger projects.

Hiring for Skills and Fit

Back in the day when I catered, I hired assistants to help wash dishes and serve. One day, my assistant Jean showed up for work with his friend Therecia. "She needs the work more than I do," he said. His implication: from that day forward, she was working for me instead of him. "It's just washing dishes," he added. "Did you tell her what the job is?" I asked of Jean. He nodded. And with that, he left.

I looked at Therecia and particularly her shoes. She wore stylish leather boots that had a one-inch heel. "Are you going to be okay in those? It's going to be a lot of standing." "They're very comfortable," she replied. It was an awkward start to her employment with me.

I was grateful to have someone in the kitchen with me. And as Jean had said, how hard is it to wash dishes? Not particularly hard, but one needs to be systematic and efficient to manage the piles of dishes that stack up very quickly. All the dishes need to be scraped first, then rinsed, then washed, then rinsed again. If you scrape, rinse, wash and rinse each plate one at a time, you'll be washing dishes all night. The fast pace of the kitchen requires a team to work in sync with each other, watching your colleagues in action and supporting their work. The same is true of farming.

On that first night, Therecia worked fast, but at some point I looked over and she was barefoot. The shoes were too uncomfortable for that kind of work. She learned quickly how to work efficiently and in sync with me, and she learned even quicker what shoes to wear. After six months, she was a tremendous asset to the business.

While I didn't have much choice in my decision to hire Therecia, I learned a lot about what makes a good employee and the skills I needed for my business.

As a business owner, you can take any warm body that comes your way (as I did), or you can make deliberate choices in the kinds of people with whom you want to work. Crafting a job description will help you articulate the skills and work-styles that are a fit for your farm. Think about these three components:

- What are the tasks?
- What are the skills and experience required?
- What is the temperament required?

Tasks

List the different duties of the person you want to hire. As a CSA manager, they may be in charge of talking with customers and managing money. Field crew employees will weed, till, harvest and package. A driver needs to load a truck, navigate city streets, deliver product and address questions with customers.

Skills

The tasks will inform the skills required. A field crew person needs to know how to drive a tractor and use farm equipment. A CSA manager needs to understand your software management system and be good at customer service. The driver needs a commercial license, people skills and the ability to lift heavy boxes.

Some jobs require customer interaction, but not all. Some require attention to detail, but not all.

Experience

Most jobs will require some level of training, such as your specific POS system or how you like to manage inventory. Some skills can be learned on the job or through training: whether it's using specific software (such as MailChimp, QuickBooks or Word) or knowing how to milk a cow or shear a lamb.

Potential hires with experience will require less training from you, the employer. This also means you will pay a higher wage for these employees. You may prefer to have young, inexperienced workers so you can train them exactly as you want them to perform their jobs. Or you may want someone who can come in, put their head down and "get 'er done."

Understanding the level of experience you desire in your employees will save headaches later.

Temperament

Just because you have the skills required for a job, doesn't mean you have the temperament for it. A computer-savvy person may have terrible interpersonal skills and may not be good at talking with customers. Further, some people are not suited to work in a fast-paced environment. You'll be spending a lot of time with your employees; and while you don't necessarily want them as friends, you want to like and respect them.

As you think about hiring, list the qualities you value in a worker, such as :

- Good with people
- Self-starter
- Flexible
- Punctual
- Likes a fast-paced environment

Not all hires necessitate an official job description. But for your own clarity and for interviews, it helps to have these listed. If you will use a job description for a public posting, you can use the Job Posting Template.

Marketing

With a growing business, marketing can no longer be an afterthought. You need to ensure you have a market for your expanded offerings and find the customers to buy your products. This requires a clear vision of your customers, what they value and how you will find them.[2]

Identify Your Target Customer

Remember, your customer is not necessarily the end-user. If you sell to restaurants, the diners at the restaurant are the ultimate end-user, but your customer is the chef/manager of the restaurant. To whom will you be selling?

Where do they live? What do they do for work? How much money do they earn? In as much detail as you can, describe your customer.

Clarify the Value of Your Products.
What is it about your products that your customers will appreciate? If you sell wholesale, they may appreciate that you now have the capacity to contract grow. If you expand into agri-tourism, your customers may value the educational opportunities for their kids. Detail the value your customers gain from your products. This will define the message that you want to communicate to them.

What Are the Habits of Your Customers?
What do they read, where do they shop? How can you find them to tell them about your product? Be as specific as possible. For retail sales customers, you may have a different strategy than for wholesale customers. By understanding their habits, you can decide the best way to reach them—whether it's by advertising in magazines, posting on Facebook, sampling your product at a local market or walking into the backdoors of restaurants.

Allocate a Budget
You know who your customers are, what's important to them and where to find them. Now you can create a strategy to communicate your value to them. With targeted advertising on Facebook and Google, you can easily target retail customers. You may also want to buy a mailing list and send an email blast and/or advertise in a local journal. For wholesale customers, your budget can be used to direct call/visit potential customers and bring samples of your product.

Key Success Factors
All this advice can be distilled into four main points to improve your chances of success in growing your business:
- Run your numbers: make sure your plan is financially viable
- Capitalize appropriately
- Develop your management skills
- Build your customer base

• • • • •

Like most things in life, we learn the most from our challenges and failures. The stories in this book are based on real farmers and real experiences; it can feel discouraging to read about all the challenges.

But there are happy endings. All the hard work of business planning pays off. As I revisited clients to hear how our work together impacted their businesses, I was heartened to know about so many successes.

When I worked with CJ and Jean from Knight Farm, they needed a loan to build a farm stand and expand their pig operations. They received the loan, built the stand and started selling their own pork products. Today their business is thriving: the farm store is bustling with customers, and they are finally paying themselves a good salary. They are thinking about further expanding into beef, and now can invest in cattle from their own savings instead of borrowing.

Brett from Even' Star Organic Farm took the lessons learned from his cost analysis to scale back his summer operation and focus on winter cropping. It was clear that the winter production was a better business model. As Brett evaluates the next phase of his farm, business planning has helped him realize that his real love, crop genetics, will be more profitable and less of a physical strain. Even' Star is poised to continue its winter production and, in the summer, a site for onsite crop breeding.

Dan from Solstice Farm discovered that smaller was better. His scaled back operations; not only has his business become more profitable, but it allowed him a better quality of life.

Proper business management doesn't solve all problems, but it positions you for success and the freedom to explore your life's passions.

Notes

1. These concepts were codified by Creative Realities.
2. For further reading about marketing, check out *Permission Marketing* by Seth Godin and *Guerilla Marketing* by Jay Conrad Levinson.

Glossary

Annual Revenue

Sometimes referred to as gross income, this is the total earnings generated through the sale of your goods and services over the period of one calendar year. This is the total amount brought in by the business without regard to operating expenses. Revenues do not include income from sources other than your primary line of business. For example, you may earn interest on a savings account or from the sale of an asset.

Assets

The textbook definition is: "Something that gives future economic benefit as the result of a past transaction." Assets are listed on the Balance Sheet.

Assets can include your home, car, land, building, large equipment, fixtures and so on. Typically, the value is greater than $1,000 and has a useful life of longer than 1 year.

When creating a balance sheet, the full current value of the asset should be declared even if you have debt associated with it. The value of the loan associated with the asset is declared separately in the liabilities section. You may also include in your list of assets: inventory and accounts receivable (money that customers owe you for products already sold). (See also: Balance Sheet, Liabilities.)

Balance Sheet

One of three financial statements most requested by investors and lenders (the other two being the income statement and cash flow statement), the balance sheet offers a summary of your business's financial position at a given point in time, usually at the end of the calendar year, detailing assets, liabilities and owner's equity. The balance sheet tells what you have (assets) and how you got it, whether it's a loan (liability) or through earned income and owner's investment (owner's equity). The owner's equity can be calculated by subtracting liabilities from total assets. In other words, this equation always holds true on a balance sheet: assets = liabilities + owner's equity. Also see video for more details (juliashanks.com/the-balance -sheet/). (See also: Assets; Liabilities.)

Business Plan

A document prepared by the entrepreneur detailing the past, present and intended future of the enterprise. It explains the business's strategy to achieve profitability and mission-related goals. The business plan serves two purposes: it helps the entrepreneur think through all details of the business, and it shows this thought and research to the investor to demonstrate that the entrepreneur will be a good steward of their money. For an outline of a business plan, see Appendix 2.

CAPEX

Shorthand for "capital expenditures," this refers to money spent on purchasing or upgrading physical assets such as equipment, buildings or machinery. Capital expenditures differ from operating expenses in that the CAPEX give long-term benefit to the business. CAPEX are depreciating assets. By contrast, routine repairs are considered operating expenses. (See also: Assets; Depreciation; Expenses.)

Capital Investments

Money invested in the business with the expectation that the investment will generate income in the future. It is used for capital expenditures as opposed to day-to-day operations or other operating expenses. Also see video for more details juliashanks .com/cpitalized-costs/.

Capital Plan

This is a list of capital expenditures a business owner intends to make in the next few years. (See also: CAPEX.) Here is a sample capital plan:

In order to grow our business to meet our goals, we expect to have the following capital expenses:

- 2009: $13,000 (Basket weeder, Planet Jr. seeder, Tractor, Hay rake, Winstrip trays
- 2010: $16,000 (Mulch layer, Waterwheel transplanter, Diesel pickup truck, High-clearance cultivating tractor
- 2011: $30,000 (Greenhouse [30′ by 96′], Large tractor with creeper gear, loader and pallet

In the following two years, we will need to raise capital for the purchase of the current farm location from the owner, which is $100,000, or an existing farm in a new location.

Cash Flow Statement

The statement of cash flows is one of the three financial statements that you will need to include in your investment package for potential investors. Cash flows refer to both inflow (cash that comes in to your business) and outflow (cash you spend on expenses, to purchase assets or pay back loans).

The cash flow statement is divided into three categories: operating, investing and financing.

- Cash flow from operations includes cash inflows from revenues and outflows for operating expenses.
- Cash flow from investing includes cash inflows from the sale of equipment or other assets and outflows from the purchase of equipment or other assets.
- Cash flow from financing includes cash inflows from grants, loans and equity investments and outflows from the repayment of debt or the distribution of dividends.

Due to the seasonality of businesses, it's important to show cash flow projections by month for the first two years of your new venture. While the overall cash flow for the year may be positive, it's important to recognize that during the slow periods it may be negative (meaning more cash is going out of the business than is coming in), and you'll need to plan for those periods. Also see video for more details juliashanks .com/cash-flow-statement/.

Collateral

An asset pledged as security for a loan. When borrowing money, the borrower provides specific property as repayment if unable to repay the debt with cash. If collateral is unavailable, a co-signer who guarantees the loan can also be used.

Cost Accounting

A method of tracking the expenses of producing goods in order to inform management decisions. Any expense that can be directly tracked to a specific product or service is tracked. It allows the entrepreneur to understand the true cost (and profitability) of each product or service and make adjustments as necessary.

Cost of Goods Sold

The price you pay for items purchased to be resold. This is the direct cost of the products. It is often abbreviated as COGS, and is sometimes referred to as Cost of Sales (COS).

Current Maturity of LTD

Long Term Debt (LTD) is any loan that is not due to be repaid in full for at least one year. For many loans, such as car loans or mortgages, a portion of the principal is due to be repaid each year. The amount of principal that is due within the next 12 months is the current maturity of LTD. (See also: Interest Expense.)

Debt Service

The amount of cash required to pay principal and interest on all loans over a certain period of time. For example, let's say you take out a loan for $50,000 loan with 5% interest rate for 5 years. The monthly payments on the loan will be $943.56. Annual Debt Service is $11,148.18 (monthly payment $943.56 × 12). (See also: Interest Expense, Principal.)

Debt Service Coverage

The amount of cash available, after paying operating expenses, to meet debt service obligations. (See also: EBIDA.)

Debt Service Coverage Ratio

Expressed as a formula: (Net Annual Income + Depreciation)/Total Annual Debt Service. Ideally, the ratio is greater than 1.2. This means you earn enough profits in your business to pay your annual debt obligations. (See also: EBIDA.)

Depreciation/Amortization

This refers to the decline in value of an asset, and the process of allocating that amount as an expense on the income statement over several years. Depreciation is used for tangible assets such as a car or equipment. Amortization is used for intangible assets like a patent. All assets, except for land, have a limited life and decline in usefulness and value over time. When an asset is purchased, the annual depreciation is calculated based on the expected life of the asset.

Expressed as a formula: Annual Depreciation = (Purchase Price – Salvage Value)/Expected Useful Life, where salvage value is the amount the asset is worth at the end of its useful life.

Each year, annual depreciation will be recorded on your income statement. By adding this non-cash expense to your income statement, you can account for the value that the equipment offers to your daily operation. In addition, you will record the "book value" of your assets on your balance sheet. The book value of the asset is calculated by subtracting the total (accumulated) deprecation from the original purchase price. For more details see the Depreciation Video at juliashanks.com/depreciation/ (See also: Assets; Balance Sheet; Net Income.)

Dividends

Distribution of net income to the business owners. This is different than salary: they don't appear on the income statement and do not have payroll taxes deducted. Dividends are listed on the statement of cash flows in the "financing section."(See also: Cash Flow Statement.)

Due Diligence

The process by which a lender or investor reviews an entrepreneur's business plan and financial projections to determine whether or not it's a sound investment.

EBIDA

Earnings Before Interest Depreciation and Amortization. Net income includes the impact of interest, depreciation and amortization on profitability; EBIDA offers a better sense of operating cash flow and the borrower's ability to repay debt. It takes into consideration a more conservative impact of income taxes on cash flow than EBITDA (Earnings Before Interest Taxes Depreciation and Amortization).

EBIDA = Net Income + Interest Expense + Depreciation Expense + Amortization Expense
    (See also: Net Income.)

Expenses

The outflow of cash or cost incurred as a result of the ordinary course of running the business, such as for the purchase of fuel and top soil, or payment to employees.

Financial Viability

In order for a business to succeed, it needs to have an effective operating plan and strategy for earning revenues and controlling expenses. If the business can operate consistently with a profit (where revenues exceed expenses), as well as pay off loans, it is considered financially viable. Often, a business will need to make an investment to improve viability. For example, a farm might install a greenhouse in order to extend the growing seasoning, thereby increasing revenues. By making this investment, the farm has improved its financial viability.

Fixed Interest Rate

An interest rate that does not change during the time that a loan is being paid back.

Gross Profit

As a formula, it is expressed as Revenue – COGS. It tells a business manager how much profit she makes on the goods sold and the money available to pay for operating and overhead expenses.

Gross Revenue

Sometimes referred to as gross income, this is the total money generated through the sale of your goods and services over the period of one calendar year. This is the total amount brought in by the business without regard to operating expenses.

Income Statement

Also referred to as the Profit and Loss, the income statement lists the company's operating revenues and expense. It demonstrates that the business can earn a profit through the course of running its business. When creating pro forma income statements for a loan application or investor package, be sure to include 3 to 5 years. The first 2 years should be detailed by month. For more details, see juliashanks .com/the-income-statement/ and juliashanks.com/multi-step-income-statement-in -depth/

Interest Expense

Noted on the income statement, interest expense is the amount of interest paid on debt owed for a specific period of time. This does not include repayment of principal. For example, you take out a loan for $50,000 loan with 5% interest rate for 5 years. The monthly payments on the loan will be $943.56. A portion of the payment is allocated to pay down the principal, and a portion is allocated to pay interest expense. At the end of the first year, you will have paid a total of $11,148.18. Of that, $2,294.98 is interest expense and $9,027.13 in paying down the principal.

- Annual Interest Expense: $2,294.98
- Annual Debt Service: $11,148.18 (monthly payment $943.56 × 12)
- Principal on the Loan at the end of the year: $40,972.87
- Amount Recorded on the Liabilities Section of your year-end Balance Sheet: $40,972.87

(See also: Debt Service; Principal.)

Liabilities

The textbook definition is "Future obligations as the result of a past transaction." Liabilities are debts that a person or business owes. They are listed on the balance sheet. The liabilities section includes loans or other debts due. This may include car loans, mortgages, credit card debts, as well as outstanding payments due to vendors and suppliers. In the case of mortgages and other loans, only include the amount currently owed, not the original amount borrowed. If you received deposits from customers for goods or services to be provided at a future time, these would also be liabilities.

Microloan

A very small, short-term loan. Historically, this has been a tool for impoverished entrepreneurs in developing countries, but is now being offered to small entrepreneurs in need of small financing solutions. In the US, microloans may be offered in amounts ranging from $500 to $50,000.

Net Income

Net income (also referred to as net profit) is total operating revenues minus total operating expenses, and is detailed on the Income Statement. Annual net income is the amount of profit earned for one calendar year. Total expenses include operating expenses, interest expense, taxes, depreciation and amortization. Expenses do not include repayment of principal on debt. (See also: Annual Revenues, Principal.)

Period

This is shorthand for "a period of time." The time frame can be a month, quarter or year.

Primary Markets

A description of where and to whom you sell the majority of your products. This may be at a farmers market or a farm stand, through a CSA, restaurants or a wholesaler.

Principal

The amount of money borrowed that is still owed on a loan. This is separate from interest expense. For example, you take out a loan for $50,000 with 5% interest rate for 5 years. The monthly payments on the loan will be $943.56. A portion of the payment is allocated to pay down the principal, and a portion is allocated to pay interest expense. At the end of the first year, you will have paid a total of $11,322.72. Of that, $2,294.98 in interest expense and $9,027.13 in paying down the principal.

- Annual Interest Expense: $2,294.98
- Annual Debt Service: $11,322.72 (monthly payment $943.56 × 12)
- Principal on the Loan at the end of the year: $40,972.87
- Amount Recorded on the Liabilities Section of your year-end Balance Sheet: $40,972.87

(See also: Debt Service; Interest Expense.)

Projections

A set of financial statements (income statement, balance sheet and statement of cash flows) prepared by the entrepreneur that outlines anticipated future financial performance. Projections are based on historical performance and assumptions of future performance. A projection is your best guess at how things will go during the year and should be supported by the details in your business plan.

Sales Channel

Sales channels are the avenues by which you sell to your customers. For a farmer, your customers may be restaurants, distributors, cooperatives, consumers. Your customers are different from the end-user, the person who consumes your product. Sales Channels can be direct to your consumers, such as at the farmers market, or indirect, such as through a distributor or cooperative.

Sales Mix

The range of products that you sell. For a dairy farmer, your sales mix could include liquid milk, cheese and ice cream. A market farmer may have a sales mix of fruits, vegetables, flowers and eggs. The sales mix can also describe the ratio of each product that you sell. As an example, 55% of your total sales come from milk, 30% comes from cheese, and 25% comes from ice cream.

SWOT Analysis (Strengths, Weaknesses, Opportunities and Threats)

This analysis addresses the business's internal strength and weakness as well as external factors that provide opportunities or pose threats to the business. When doing a SWOT analysis, it's important to be realistic about your strengths and weaknesses.

Examples of strengths include:
- Our low employee turnover leads to higher labor efficiency.
- Ability to meet demands of the growing market.
- Suzy Q has strong personal relationships with many of her customers and local businesses in the areas where she works. These strong relationships have created many loyal customers and positive word of mouth.

Examples of weaknesses include:
- Lack of equipment means the farm does not utilize labor as efficiently as it could.
- Do not own land.
- Dependency on other farms to produce our value-added products.

Examples of opportunities:
- Growing public interest in local and organic products.
- Several new farmers markets are opening in our region providing more venues to sell our products.

Examples of threats:
- US economy is unstable, and consumers may be less willing to pay a premium for local and organic products.
- The severe weather (rains of last year, cold winter) diminishes yields.
- The price of grain is rising which will directly increase the cost of animal feed.

Treasury Rate

The Treasury rate refers to the current interest rate that investors earn on debt securities issued by the US Treasury. US Treasury securities are considered to be the safest debt investment. The interest rate paid on Treasuries sets the benchmark for all other loans.

Venture

By definition, a venture is a risky undertaking or journey. In entrepreneurial terms, it's the beginning of a new business or business extension.

Yearly Debt Service

The amount that you pay each year in principal and interest for outstanding loans. (See also: Debt Service; Interest Expense; Principal)

Sample Income Statement Accounts

Expenses that are *italicized* are usually considered variable expenses (as opposed to fixed) and increase as sales increase.

Revenues
Sales
CSA
Farmers Market
      Farm Stand
      Wholesale
   Total Revenues

Cost of Goods Sold
   Animal Purchases
   Produce for Resale
   Farm Store/Farm Store Products for Resale
Total COGS

Gross Profit

Expenses
   Direct Operating
      Bedding
      Chemicals
      Farm Stand Supplies
      Farmers Market Supplies
      Feed and Supplements
      Fertilizer, Lime and Soil Amendments
      Transportation Costs
      Gasoline, Fuel, and Oil
      Small Equipment
      Equipment Rental

Other Supplies
Packaging
Processing/Slaughter
Row Cover
Scales
Seeds and Plants
Shearings/Shavings
Top Soil/Potting Soil
Veterinary/Medicine
Total Direct Operating

Payroll
Employee Benefits
Health Insurance
Hourly Labor Expense
Salary Expense
Payroll Service
Payroll Tax Expense
Worker's Comp Insurance
Unemployment Insurance
Total Payroll

General and Administrative
Auto Insurance
Automobile Expenses
Bank Service Charges
Cleaning Supplies
Computer and Printer
Crop Insurance
Dues and Subscriptions
Insurance - Liability
Internet
Licenses and Permits
Office Supplies
Postage and Shipping
Professional Fees
Registration
Telephone
Vehicle Lease Payment
Total General and Administrative

Advertising and Promotion
    Advertising Expense
    Promotional Items
    Website Maintenance
Total Advertising and Promotion

Repairs and Maintenance
    Animal Housing
    Automobile
    Buildings
    Coolers/Refrigerators
    Equipment
    Fencing
    Greenhouse(s)
    Irrigation
Total Repairs and Maintenance

Occupancy
    Rent
    Utilities
    Gas and Electric
    Water
    Property Tax
Total Occupancy

Total Operating Expenses

Net Operating Income

Other Income
    Grant Income
    Interest Income
    Rental Income
Other Expenses
    Depreciation
Income Taxes
    Interest Expense
Total Other Income/Expenses

Net Income

The Business Plan

The business plan is a great way to organize the swirl of ideas about your new business, and an opportunity to make sure you think through all the different components of your new business or enterprise.

This outline is intended to coach you through the process of writing a business plan. **Each header** indicates the heading of each section that should be included. Under the headers *in italics* is a basic overview of each section and what it includes. Finally, **we list all the details** that should be included in your business plan, with questions to help you think through how you want to structure your business.

It is not required that you answer every question; some may not make sense for your business. It's better to leave out a section than to include language that sounds contrived.

The business plan is the narrative that accompanies your financial projections. As you work through the financials and plan, your vision will evolve.

Executive Summary
This section should be written last because it's read first. It's an overview of your farm concept, market opportunity, operations overview and any requested financing. It should be no longer than one page.
- Farm name
- Concept, including products and philosophy
- Location
- Target customers
- Overview of financial projections
- Amount of money needed to be raised and expected returns on investment

Farm/Business Description
A 2 to 3 page basic overview of the farm.

Farm Name
What was the inspiration for the name and concept?

Business Entity
- Legal form of ownership. Are you an LLC, Sole Proprietor, C-Corp, B-Corp
- Why have you selected this form?

Location
- Where is the farm located or where do you plan to locate the farm? Why did you pick that location?

Size of Farm
- Total number of acres on the property; total acres in production or planned for production?
- Number of acres used or planned to be used for different enterprises (i.e. vegetables, livestock, poultry, cut flowers, etc.)?
- Infrastructure and its use or planned use?
    - Is there a barn? Greenhouse? If you don't yet have a farm site, what infrastructure will be necessary?

Mission Statement
- What is the purpose of your company?
- What is your philosophy of doing business?
- What are your goals?

Type of Products
- What do you/will you grow/raise?
- Influences? Specialty items?
- Value-added products?
- What differentiates your products from other farms? If you are not yet in business, what do you see as your potential distinguishing factors?

Sales Channels
Where do you/will you sell your product? For each sales channel provide a few sentences about why the sales channel is a good avenue for you. Some sales channels include:
- CSAs
- Farmers Markets
- Wholesale
- Farm stand
- Co-ops
- Where else?

Current Status of Development
- Are you already farming? If not,
  - Has a site been selected?
  - Have you built infrastructure?
  - Has your team (field crew, management and business advsiors) been solidified?
- If so:
  - How many years have you been farming at the current location?
  - Are you an early-stage business or well-established? That is, how many years have you been in operation?

Future Plans
- If you are already farming, do you plan on expanding the farm? When or at what stage?
- Would you expand product offerings? To what?
- What are your sales/revenue goals?

Industry Analysis
What is going on in the farm and food industry? How do these events affect trends, markets and sales? How do these trends impact your business? What are the current opportunities? This section should demonstrate an understanding of the agriculture industry as it relates to your business.

This section doesn't need to be more than a page or two. If you are proposing less common farm-income sources, such as agri-tainment or value-added products, you may want to expand on why the time is right for you to launch or expand your business based on industry trends.

Current State of the Agriculture Industry and Resulting Trends
- For the type of products you are/will be offering
- For the industry in general

Resources
localharvest.org/
smallfarmtoday.com/
dtnprogressivefarmer.com/dtnag/
newventureadvisors.net/marketsizer.php
ers.usda.gov/topics/farm-economy.aspx, rodaleinstitute.org/farm/organic-price
  -report/
marketnews.usda.gov/portal/fv

SWOT Analysis

SWOT stands for Strengths, Weaknesses, Opportunities and Threats. A SWOT analysis addresses the business's internal strengths and weaknesses, as well as external factors that provide opportunities or pose threats to the business.

The weaknesses and threats provide an opportunity for the entrepreneur to hone operations and position it for growth.

When doing a SWOT analysis, it's important to be realistic about your strengths and weaknesses.

Examples of strengths:

- Low employee turnover on the farm leads to higher labor efficiency
- The farm's ability to meet demands of the growing market

Examples of weaknesses:

- Lack of equipment means the farm does not utilize labor as efficiently as it could
- Do not own land
- Dependency on other farms to produce our value-added products

Examples of opportunities

- Growing public interest in local and organic products
- Several new farmers markets are opening in our region providing more venues to sell our products

Examples of threats

- US economy is unstable, and consumers may be less willing to pay a premium for local and organic products.
- Severe weather (flooding, early/late frosts) could diminish yields.
- The price of grain is rising which will directly increase the cost of animal feed.
- Avian flu is raising the cost of eggs.

As you think about your strengths, weaknesses, opportunities and threats, consider the following:

- What skills do you have that make you qualified to start and run your businesses?
- What resources are available to you that will help your business succeed?
- What are current opportunities to collaborate with other businesses or capitalize on current trends? Some examples:
  - Can you partner with another farmer to sell products through his farm store?
  - Can you partner with a caterer to create value-added products?
- What trends do you foresee and how will they impact your business? How will you address future trends?
- What measures will you put in place to mitigate threats?

Products and Related Services
A more detailed description of your products and services.

Service and Product Offerings
- Products might include root crops, leafy greens, pasture-raised beef, preserved/canned goods, etc.
- Services might include a CSA, custom orders, farm stand hours, etc.

Production
- If you are already in operation, are all of the products you sell produced on-farm or are some purchased from other farms/food producers?
- If you aren't yet operating, how much of each of your product offerings will be produced on-farm versus purchased from other producers?
- What growing practices do you use? Organic? IPM? Biodynamic?

Future Opportunities
- What opportunities exist to grow the business? Expand into additional enterprises? Offer your products though additional sales channels?

Marketing Plan and Strategy
Approximately 2 pages that outline who your customers are/will be, based on location and demographics, and how you will reach them. Include information about branding, communication strategies, calendar and budget for marketing efforts.

Market Location and Customers
- Who are/will be your target customers based on your farm's planned location?

General Demographics of Your Planned Location
- Residents
- Businesses
- Schools and universities
- Proximity to city/population center
- What do the surrounding towns look like? Do you/will you draw customers from them?

Customer Profile
Who are/will be your customers, and what do they do? You should have more than one.
- Are they families or individuals?
- Are they businesses and institutions?
- Where are they located?
- How old are they?
- What is their education?

- What is their lifestyle?
  - What is their median income
  - Do they seek value pricing or quality pricing?

What Is/Will Be Attractive About Your Business to Your Target Customer Profiles?
- Price point
- Lifestyle choices
- What is/will be the clincher for a customer to make a purchasing decision? Convenience? Destination? Comparison?
- How do you/will you communicate your brand features to your potential customers?

Market Trends and the Future
- Where do you see your segment of the agriculture industry going? How will your concept adapt and change with trends?

The Competition
Farms compete with "share of stomach." Consumers can only eat so much. The competition can be farms located nearby, farms in the region with a similar concept or other options for food purchases, like grocery stores, that may deter potential customers from purchasing from your farm.
- Who is/ will be your competition? List direct and indirect competitors (i.e., another farm, grocery store or wholesale distributor).
- What are/will be the features of your farm? How do they compare with the competition?
- Why do/will customers choose your farm's products over the competition?

Market Penetration
- If you are not yet in operation:
  - How will you enter the marketplace? How will you let your customers know you're there?
  - What sort of budget do you have for marketing?
  - How will you track whether marketing strategies are effective?
- If your farm is already in operation:
  - How will you expand your market penetration either with existing or new product offerings?
  - What is your marketing budget?
  - How will you track whether these marketing strategies are effective?

Ongoing Strategy

- Regardless of the current stage of your farm business, how will you continue to keep current and potential customers engaged and interested in your farm in the future?

Future Plans and Strategic Opportunities

- If you are not yet in operation, how would you plan to grow your business in the future? (e.g., appealing to a larger customer base, selling your products more widely)
- What opportunities for partnership or expansion do you anticipate? (e.g., selling products for other farms or vice versa, opening a brick and mortar store/farm stand, sales to local institutions)

Operations

What is the basic outline of your farm's operation? How much physical space do you have or will you need? What kind of infrastructure/buildings do you have or will you need?

- Tillable acres
- Pasture acres
- Wash area
- Cool storage
- Equipment storage
- Offices
- Power and other utilities
- Restrooms
- Parking
- What else?

Production Techniques

- At what stage do you/will you start the production of your goods? (e.g., start plants from seed or buy transplants? Birth livestock or buy feeder animals?)
- Do you/will you mostly rely on manual tools or mechanized equipment? (e.g., planting, weed management, irrigation)
- Do you/will you use chemicals, fertilizers, or other similar products?

Seasons and Hours of Operation

- What days do you/will you and your staff work?
- Is any part of your operation closed any day? If you are not yet operating, what kind of operating schedule do you anticipate?
- What months of the year do you/will you produce which products?

Employee Hiring, Training and Development
- If you are already in operation:
  - Do you have employees? How many? What skills do they have?
  - When you need to hire, where will you find new employees?
  - Do you have an employee training program?
  - What sort of on-going training, development, and benefits do you offer to your employees?
- If you are not yet in operation:
  - Will you hire employees? How many? What skills will they need? Where will you find them?
  - When you hire, will you put employees through a training program?
  - What sort of on-going training, development, and benefits will you offer your employees?

Systems and Controls
- What sorts of systems do you/will you have in place to ensure quality control?
- What sorts of systems do you/will you have in place to ensure a quality customer experience?
- What systems are/will be in place to control costs?
- What systems are/will be in place to manage inventory?
- What are/will be your credit and delivery policies?

Legal Environment
- What are the licensing and permit requirements for your farm business?
- What are the health, workplace or environmental regulations with which you need to comply?
- What are the zoning or building code requirements that apply to your business?
- What insurance do you/will you require?

Development Plan
- If you are not yet operating:
  - When do you anticipate selling your first products?
  - When do you want to expand the farm's operations, or otherwise grow your business?
  - How long will it take between getting the funding to opening day? What are the milestones?

Management and Organization
Leadership
- What makes you uniquely qualified to run this business? Why should investors or lenders give you money?

- How is/will the company be organized? What does/will your organization chart look like?
- What is/will be your management structure and style?
- What is/will be the ownership structure?

Key Employees and Principals
- For each *key* position (not all), detail:
  - What are/will be their duties and responsibilities?
  - What unique skills do/will they bring to the venture?
  - If you already have a team in place or if you know who your team members will be, include short bios of each key employee and add full resumes in the appendix.

Professional and Advisory Support
- Do you/will you have a board of directors, management advisory board, or mentors/key advisors?
- Do you/will you partner with consultants and professional support resources? What roles do they/will they play?
- What additions to the management team do you plan? When?

Critical Risks

There are risks in every business. While you can't anticipate everything that will go wrong, something will go wrong. The more you can think through in advance, the better chance you have to survive the challenges. What could get in the way of your being successful either in starting your business or growing it? If you are already in operation, what systems/ policies do you already have in place to manage these risks?

- Are there competitive threats you're aware of?
- What unexpected costs could occur? How will you handle them?
- What disruptions in your management could occur? How will you handle them?
- How might weather-related events threaten your production or distribution? How will you mitigate these risks?
- What else?

Offering

If your business plan is written with the goal of obtaining financing, include this section. In 1 page or less, tell the readers what sort of investment you need to launch or grow your business.

- How much money do you need to launch or grow your farm business?
- What will the funds be used for?
  - Capital improvements?
  - Working capital?

- What type of financing are you seeking?
  - Debt or equity?
  - At what interest rate? If equity, what sort of return are you offering investors?

Financial Plan
Approximately 1 page of text plus the financial statements, which are commonly attached at the end of the written plan.
- Detailed financial assumptions
  - What are your sales projections based upon?
  - What are your expense projections based upon? It's important to focus on your biggest expenses, which will be direct operating expenses and labor.
  - How did you determine how much money you need to raise?
- Pro forma financial statements for 5 years, the first 2 years by month; years 3–5 can be an annual summary.
  - Income statement
  - Balance sheet
  - Statement of cash flows

Long-term Development
What are your long-term goals for the business?

Goals
- List specific goals you have for your business. They could be revenue goals, profitability goals, or market reach within a certain number of years.

Milestones
- What milestones do you want to measure/track?
- How many customers (CSA shares, wholesale customers, institutional partners, etc.) do you want to have? By when?
- How much revenue? By when?
- How much profit? What percentage of revenue would that comprise?

Sample Chart of Accounts
for Livestock and Vegetable Farms

Note: Colons separate the parent accounts from the sub-accounts. Account names in **bold** are parent accounts.

Livestock Farm COA

Account Name	Account Type
Sales	Income
Sales:Animals	Income
Sales:CSA	Income
Sales:Farmers Markets	Income
Sales:Farm Stand	Income
Sales:Wholesale	Income
Cost of Goods Sold	Cost of Goods Sold
Cost of Goods Sold:Animal Purchases	Cost of Goods Sold
Cost of Goods Sold:Produce for Resale	Cost of Goods Sold
Cost of Goods Sold:Dry Goods for Resale	Cost of Goods Sold
Direct Operating	Expense
Direct Operating:Bedding	Expense
Direct Operating:Chemicals	Expense
Direct Operating:Feed and Supplements	Expense
Direct Operating:Fertilizer and Lime	Expense
Direct Operating:Freight and Transport	Expense
Direct Operating:Freight and Transport:Gasoline, Fuel, and Oil	Expense
Direct Operating:Equipment	Expense
Direct Operating:Equipment Rental	Expense
Direct Operating:Other Supplies	Expense
Direct Operating:Packaging	Expense
Direct Operating:Processing/Slaughter	Expense
Direct Operating:Row Cover	Expense
Direct Operating:Scale	Expense

Account Name	Account Type
Direct Operating:Seeds and Plants	Expense
Direct Operating:Shearings/Shavings	Expense
Direct Operating:Top Soil/Potting Soil	Expense
Direct Operating:Veterinary/Medicine	Expense
Payroll	Expense
Payroll:Employee Benefits	Expense
Payroll:Health Insurance	Expense
Payroll:Hourly Labor Expense	Expense
Payroll:Salary Expense	Expense
Payroll:Payroll Service	Expense
Payroll:Payroll Tax Expense	Expense
Payroll:Worker's Comp	Expense
Payroll: Unemployment Insurance	Expense
General and Administrative	Expense
General and Administrative:Automobile Expenses	Expense
General and Administrative:Automobile Expenses:Vehicle Payment	Expense
General and Administrative:Automobile Expenses:Registration	Expense
General and Administrative:Automobile Expenses:Auto Insurance	Expense
General and Administrative:Bank Service Charges	Expense
General and Administrative:Computer and Printer	Expense
General and Administrative:Dues and Subscriptions	Expense
General and Administrative:Insurance	Expense
General and Administrative:Insurance:Liability Insurance	Expense
General and Administrative:Insurance:Crop Insurance	Expense
General and Administrative:Internet	Expense
General and Administrative:Licenses and Permits	Expense
General and Administrative:Office Supplies	Expense
General and Administrative:Postage and Shipping	Expense
General and Administrative:Professional Fees	Expense
General and Administrative:Telephone	Expense
Advertising and Promotion	Expense
Advertising and Promotion:Advertising Expense	Expense
Advertising and Promotion:Promotional Items	Expense
Advertising and Promotion:Website Maintenance	Expense
Repairs and Maintenance	Expense
Repairs and Maintenance:Animal Housing	Expense
Repairs and Maintenance:Automobile	Expense
Repairs and Maintenance:Buildings	Expense

Account Name	Account Type
Repairs and Maintenance:Coolers/Refrigeration	Expense
Repairs and Maintenance:Equipment	Expense
Repairs and Maintenance:Fencing	Expense
Repairs and Maintenance:Greenhouse(s)	Expense
Repairs and Maintenance:Irrigation	Expense
Occupancy	Expense
Occupancy:Rent	Expense
Occupancy:Utilities	Expense
Occupancy:Utilities:Gas and Electric	Expense
Occupancy:Utilities:Water	Expense
Interest Expenses	Other Expense
Total Taxes	Other Expense
Other Income	Other Income
Operating Account	Bank
Payroll Account	Bank
Savings Account	Bank
Cash Drawer	Bank
Credit Card	Credit Card
Line of Credit	Other Current Liability
Gift Certificates	Other Current Liability
Payroll Tax Payable (Employee Contriubtion)	Other Current Liability
Prepayments from Customers	Other Current Liability
Daily Sales Receipts	Other Current Asset
Mortgage	Long Term Liability
Vehicle Loan	Long Term Liability
Equipment Loan	Long Term Liability

Vegetable Farm COA

Account Name	Account Type
Sales	Income
Sales:CSA	Income
Sales:Farmers Markets	Income
Sales:Farm Stand	Income
Sales:Wholesale	Income
Cost of Goods Sold	Cost of Goods Sold
Cost of Goods Sold:Produce for Resale	Cost of Goods Sold
Cost of Goods Sold:Dry Goods for Resale	Cost of Goods Sold

Account Name	Account Type
Direct Operating	Expense
Direct Operating:Chemicals	Expense
Direct Operating:Fertilizer and Lime	Expense
Direct Operating:Freight and Transport	Expense
Direct Operating:Freight and Transport:Gasoline, Fuel, and Oil	Expense
Direct Operating:Equipment	Expense
Direct Operating:Equipment Rental	Expense
Direct Operating:Other Supplies	Expense
Direct Operating:Packaging	Expense
Direct Operating:Row Cover	Expense
Direct Operating:Scale	Expense
Direct Operating:Seeds and Plants	Expense
Direct Operating:Shearings/Shavings	Expense
Direct Operating:Top Soil/Potting Soil	Expense
Payroll	Expense
Payroll:Employee Benefits	Expense
Payroll:Health Insurance	Expense
Payroll:Hourly Labor Expense	Expense
Payroll:Salary Expense	Expense
Payroll:Payroll Service	Expense
Payroll:Payroll Tax Expense	Expense
Payroll:Worker's Comp	Expense
Payroll: Unemployment Insurance	Expense
General and Administrative	Expense
General and Administrative:Automobile Expenses	Expense
General and Administrative:Automobile Expenses:Vehicle Payment	Expense
General and Administrative:Automobile Expenses:Registration	Expense
General and Administrative:Automobile Expenses:Auto Insurance	Expense
General and Administrative:Bank Service Charges	Expense
General and Administrative:Computer and Printer	Expense
General and Administrative:Dues and Subscriptions	Expense
General and Administrative:Insurance	Expense
General and Administrative:Insurance:Liability Insurance	Expense
General and Administrative:Insurance:Crop Insurance	Expense
General and Administrative:Internet	Expense
General and Administrative:Licenses and Permits	Expense
General and Administrative:Office Supplies	Expense
General and Administrative:Postage and Shipping	Expense
General and Administrative:Professional Fees	Expense
General and Administrative:Telephone	Expense

Account Name	Account Type
Advertising and Promotion	Expense
Advertising and Promotion:Advertising Expense	Expense
Advertising and Promotion:Promotional Items	Expense
Advertising and Promotion:Website Maintenance	Expense
Repairs and Maintenance	Expense
Repairs and Maintenance:Automobile	Expense
Repairs and Maintenance:Buildings	Expense
Repairs and Maintenance:Coolers/Refrigeration	Expense
Repairs and Maintenance:Equipment	Expense
Repairs and Maintenance:Fencing	Expense
Repairs and Maintenance:Greenhouse(s)	Expense
Repairs and Maintenance:Irrigation	Expense
Occupancy	Expense
Occupancy:Rent	Expense
Occupancy:Utilities	Expense
Occupancy:Utilities:Gas and Electric	Expense
Occupancy:Utilities:Water	Expense
Interest Expenses	Other Expense
Total Taxes	Other Expense
Other Income	Other Income
Operating Account	Bank
Payroll Account	Bank
Savings Account	Bank
Cash Drawer	Bank
Credit Card	Credit Card
Line of Credit	Other Current Liability
Gift Certificates	Other Current Liability
Payroll Tax Payable (Employee Contriubtion)	Other Current Liability
Prepayment from Customers	Other Current Liability
Farm Store Inventory for Resale	Other Current Asset
Daily Sales Receipts	Other Current Asset
Mortgage	Long Term Liability
Vehicle Loan	Long Term Liability
Equipment Loan	Long Term Liability

Financial Projections

Congratulations! If you're here, it means that you've decided to tackle financial projections. Don't try to get through this in one session. It can take several weeks (and a few bottles of aspirin). For non- numbers people, it's a lot of numbers!

The 9 steps listed match the 9 steps in Chapter 4: The Business Planning Process. It may be helpful to refer to that chapter as you go through creating your projections.

As we work through the projections, we will use two real farms' projections to illustrate how to pull the numbers together.

- Stone Hill Farm: Kristen is borrowing $15,000 to purchase a box truck, a greenhouse and a few pieces of farm equipment.
- Knight Farm: The Knight family is applying for a $75,000 grant to build a farm store to sell their own vegetables and pork.

1. Start with Historicals (to inform assumptions)

When creating projections about future business performance, the best starting place is the past. What did your business do last year—what did you sell, what were your expenses, how much was payroll, and so on. The future numbers will, no doubt, differ; but you have a baseline to begin.

Businesses already in operation will ideally have a well maintained QuickBooks file (see Chapter 5: Getting Started in QuickBooks). Perhaps you tracked them on an Excel spreadsheet, or have your Schedule Fs from your tax returns.

When launching a new venture, there are no historical numbers to pull from. But there are numbers out there; others have forged paths before you to start farm businesses, so there is precedence. If you are currently working on a farm, you can ask your boss about some of her numbers. Or find a mentor who has run a similar business to get a better understanding. We'll go into more detail in the next section.

Kristen started Stone Hill Farm after managing a neighboring farm. With a 30-year lease on the land, she comfortably started building infrastructure including hoophouses and a packing shed. She sold her produce through a CSA and three farmers markets. She was generating profits, but not enough to support her family without off-farm income. To create a financially sustainable business, she needed

TABLE A4.1

	2010	2011	2012
Revenue			
Crop Sales	33,862	49,298	148,662
Resold Crops	29,436	38,886	—
Holiday Sale	1,330	—	—
Total Revenue	64,628	88,184	148,662
Cost of Sales	16,566	24,026	21,768
Gross Profit	48,062	64,158	126,894
Direct Operating Expenses			
Booth Fees	1,380	921	2,155
Chemicals	632	86	—
Equipment	731	—	1,983
Fertilizer and Lime	634	1,114	1,045
Mulch			5,050
Pest Control			1,381
Rent or Lease: Vehicle, Machinery and Equipment	—	166	
Seeds and Plants	1,936	1,356	3,594
Soil Tests	—		
Supplies	2,959	4,620	8,889
Total Direct Operating	8,272	8,263	24,097
Payroll			
Labor Hire	—	—	42,283
Taxes: Payroll	—	—	
Worker's Comp + Disability	—	—	1,871
Total Payroll	—	—	44,154
General and Administrative			
Accounting Services	—	—	265
Advertising	306	69	1,081
Bank Fees	—	—	440
Insurance (excluding health)	360	1,276	749
Continuing Education	744	29	30
Meals and Entertainment	60	34	164
Office Supply	1,244	222	772
Other Expense	—	1,848	—
Permits and Licenses	266	110	10
Professional Fees	—	125	450
Shipping	156	32	—
Storage and Warehousing	2,659	—	—
Total General and Administrative	5,795	3,745	3,961
Repairs and Maintenance			
Car and Truck	3,216	5,390	3,847
Gasoline	—	5,238	4,837
Repairs and Maintenance	1,342	1,806	292
Tools	731	306	1,347
Total Repairs and Maintenance	5,289	12,740	10,323
Occupancy			
Rent or Lease: Other	—	1,226	3,325
Utilities	528	—	3,000
Total Occupancy	528	1,226	6,325
Total Operating Expenses	19,884	25,974	88,860
Operating Income	28,178	38,184	38,034
Interest			664
Depreciation	16,239	10,549	12,549
Income Before Taxes	11,939	27,635	24,821
Taxes	48	391	652
Net Income	11,891	27,244	24,169

to increase sales and improve production efficiencies. To accomplish this, she identified several investments: a greenhouse to increase production, a tractor to improve efficiencies, a walk-in cooler to improve post-harvest quality and a box truck for deliveries.

Kristen had three years of tax returns. We reorganized them for clarity, and used those as a starting point for her financial projections (see Table A4.1).

For businesses that have been in operation for more than one year, like Stone Hill, the expenses change year to year. One year, Kristen spent $1,380 in farmers market fees, the next year it was $921; seeds and plants expense decreased in her second year, and then increased in her third year.

To give these numbers perspective, look at each expense as a percentage of sales (see Table A4.2).

How to Think About Historicals

Is the business profitable? In the case of Stone Hill, not only did she earn a profit, but the profit as a percentage of sales was an impressive 16.3% in her 3rd year. A non-profitable business can take this time to evaluate where expenses can be reduced and/or revenues increased to improve profitability.

How was the business financed? With no interest expense until the 3rd year, these historicals suggest that Kristen was able to launch her business with savings (rather than taking a loan). If she decides to borrow more money to grow, the interest expense will cut into her profits.

Did the farmer take a salary? It looks like Kristen was working by herself in the first two years as she had no payroll expense. Further, she had a positive cash flow from operations from which she could draw a salary ($28,130 in year 1).[1] This is pretty good. If she can maintain profits at 16.3% (as she did in year 3) while growing revenue, she will be in good shape.

How did sales grow year to year? Is this a sustainable growth rate? Or will revenue in the future grow more slowly? Kristen's revenue grew quite aggressively (close to 50% from the first year and 200% from the second). During the first few years of business, this kind of growth is possible (especially when growing from $25,000 to $50,000), but this rate is not sustainable. Eventually, you will saturate your market.

Do any expenses seem out of line? For Kristen, supplies seem to be a big expense, at 6% of revenue. "Supplies" is a broad description and could include many things. It's worth digging deeper to understand why Kristen is spending so much money here, whether it seems appropriate, and if this will continue.

Review each expense: is it a variable cost or a fixed cost?[2] Most of the direct operating expenses are variable; as revenue increases, so will the expense. Payroll will also be variable. farmers market fees are more stable.

Do you have items that you re-sell? In a farm store, you may purchase other farmers' produce or value-added products. What do you pay for those items? How much do you pay as a percentage of what you sell them for?

This review of the numbers helps as you outline the assumptions; it's also an opportunity to evaluate how the day-to-day management of your business is reflected in the numbers and how you might want to adjust.

2. List Operating Assumptions

Whether or not you have historical numbers, you need to list assumptions about how you think your future operations will go: what you think you will earn in revenue and what expenses you'll incur.

TABLE A4.2

	2010	2011	2012	2010	2011	2012
Revenue						
Crop Sales	33,862	49,298	148,662	52.4%	55.9%	100.0%
Resold Crops	29,436	38,886	—	45.5%	44.1%	0.0%
Holiday Sale	1,330	—	—	2.1%	0.0%	0.0%
Total Revenue	64,628	88,184	148,662	100.0%	100.0%	100.0%
Cost of Sales	16,566	24,026	21,768	25.6%	27.2%	14.6%
Gross Profit	48,062	64,158	126,894	74.4%	72.8%	85.4%
Direct Operating Expenses						
Booth Fees	1,380	921	2,155	2.1%	1.0%	1.4%
Chemicals	632	86	—	1.0%	0.1%	0.0%
Equipment	731	—	1,983	1.1%	0.0%	1.3%
Fertilizer and Lime	634	1,114	1,045	1.0%	1.3%	0.7%
Mulch	—	—	5,050	0.0%	0.0%	3.4%
Pest Control	—	—	1,381	0.0%	0.0%	0.9%
Rent or Lease: Vehicle, Machinery and Equipment	—	166	—	0.0%	0.2%	0.0%
Seeds and Plants	1,936	1,356	3,594	3.0%	1.5%	2.4%
Soil Tests	—	—	—	0.0%	0.0%	0.0%
Supplies	2,959	4,620	8,889	4.6%	5.2%	6.0%
Total Direct Operating	8,272	8,263	24,097	12.8%	9.4%	16.2%
Payroll						
Labor Hire	—	—	42,283	0.0%	0.0%	28.4%
Taxes: Payroll	—	—	—	0.0%	0.0%	0.0%
Worker's Comp + Disability	—	—	1,871	0.0%	0.0%	1.3%
Total Payroll	—	—	44,154	0.0%	0.0%	29.7%
General and Administrative						
Accounting Services	—	—	265	0.0%	0.0%	0.2%
Advertising	306	69	1,081	0.5%	0.1%	0.7%
Bank Fees	—	—	440	0.0%	0.0%	0.3%
Insurance (excluding health)	360	1,276	749	0.6%	1.4%	0.5%
Continuing Education	744	29	30	1.2%	0.0%	0.0%
Meals and Entertainment	60	34	164	0.1%	0.0%	0.1%
Office Supply	1,244	222	772	1.9%	0.3%	0.5%
Other Expense	—	1,848	—	0.0%	2.1%	0.0%
Permits and Licenses	266	110	10	0.4%	0.1%	0.0%
Professional Fees	—	125	450	0.0%	0.1%	0.3%
Shipping	156	32	—	0.2%	0.0%	0.0%
Storage and Warehousing	2,659	—	—	4.1%	0.0%	0.0%
Total General and Administrative	5,795	3,745	3,961	9.0%	4.2%	2.7%
Repairs and Maintenance						
Car and Truck	3,216	5,390	3,847	5.0%	6.1%	2.6%
Gasoline	—	5,238	4,837	0.0%	5.9%	3.3%
Repairs and Maintenance	1,342	1,806	292	2.1%	2.0%	0.2%
Tools	731	306	1,347	1.1%	0.3%	0.9%
Total Repairs and Maintenance	5,289	12,740	10,323	8.2%	14.4%	6.9%
Occupancy						
Rent or Lease: Other	—	1,226	3,325	0.0%	1.4%	2.2%
Utilities	528	—	3,000	0.8%	0.0%	2.0%
Total Occupancy	528	1,226	6,325	0.8%	1.4%	4.3%
Total Operating Expenses	19,884	25,974	88,860	30.8%	29.5%	59.8%
Operating Income	28,178	38,184	38,034	43.6%	43.3%	25.6%
Interest			664	0.0%	0.0%	0.4%
Depreciation	16,239	10,549	12,549	25.1%	12.0%	8.4%
Income Before Taxes	11,939	27,635	24,821	18.5%	31.3%	16.7%
Taxes	48	391	652	0.1%	0.4%	0.4%
Net Income	11,891	27,244	24,169	18.4%	30.9%	16.3%

Forecast Sales

There are two ways to create a sales forecast: start with your revenue goals (top-down) or start with specific production goals (bottom-up). If you have historical sales information about your business, you can use that as a starting point: estimate how you want the business to grow and evolve in the coming years.

Top-down Sales Projections

For a farm that has been in business for a while, it can be more efficient to start with the total amount that you want to sell (or earn) in a year; let's say it's $100,000. From there, break the number down further—how much money do you anticipate earning from different revenue streams: from, say, attending farmers markets or selling wholesale?

If you decide that 75% of your revenue will come from farmers markets and 25% from wholesale, then you need to earn $75,000 from the farmers markets and $25,000 from wholesale.

- The typical farmers market runs for 30 weeks, so that's about $2500 per week.
- If you can earn $1,250 on the average market day, then you'll need to attend 2 markets per week.
- The growing season is about 26–30 weeks. That translates to about $900 per week in wholesale.
- If the average wholesale customer orders $200 worth of produce, then you need 4–5 customers.

For Kristen, she grew her sales an average of 45% in the second year and 200% in the third year. While this was an impressive growth rate, she didn't think it was sustainable long-term. Overall, 20% revenue growth seemed more reasonable. She looked deeper at each sales channel. She felt she still had room for considerable growth with the CSA market and wholesale, but would then grow more slowly. With ideas on how much she could grow, she set sales targets for the CSA, farmers markets, wholesale and other categories.

For the first year of her business plan (2013), she set specific sales targets for each category. For the subsequent years, she estimated a growth rate from the previous year.

	Projected Growth	
	2014	2015
CSA	7.0%	8.0%
Farmers Markets	7.0%	5.0%
Wholesale	20.0%	15.0%
Direct Sales	5.0%	3.0%
XMAS	10.0%	5.0%

	2013	2014	2015
CSA	20,000	21,400	23,112
Farmers Markets	141,000	150,870	158,414
Wholesale	12,000	14,400	16,560
Seedlings	2,500	2,625	2,704
XMAS	2,500	2,750	2,888
Total Sales	178,000	192,045	2,888

Looking at the 2013 projections, $20,000 in CSA revenue translated into 34 shares at $625 each; $150,870 in farmers market sales translated to $1,700 per market in revenue. To meet the wholesale targets, Kristen needed to sell $130 per week to each of 3 to 5 customers. She now has concrete sales targets. The seedling sale and Christmas sale were small one-time events.

After you create your annual sales projection for each channel, you can then estimate how much you will earn each month. This will also help as you plan for cash flow.

Here are Kristen's sales projections broken out by month for her first year. She started with the total, and then distributed across the year based on her best guess.

	Jan	Feb	Mar	Apr	May	Jun	Jul	Aug	Sep	Oct	Nov	Dec	2013
Revenue													
CSA	5,000	5,000	5,000	5,000									20,000
Farmers Markets						20,000	23,000	32,000	29,000	25,000	13,000	7,000	149,000
Wholesale				500	500	1,000	1,500	3,000	2,000	2,000	1,000	500	12,000
Seedlings				500	2,000								2,500
XMAS												2,500	2,500
Total (cash) Revenue	5,000	5,000	5,000	6,000	2,500	21,000	24,500	35,000	31,000	27,000	14,000	10,000	186,000

Bottom-up Sales Projections

Especially in the first year of a new business, it's easier to start from the bottom up with sales projections.

- How many wholesale customers do you want? How much do you expect them to order each week? How long is your season?
- For how many CSA subscribers do you think you can grow produce?
- How many dairy cows will you have? How many gallons of milk can they produce in a day? A week? A month?

Richard Knight started farming in the 1960s growing hay and raising feeder pigs. On 100 acres, he grew his business and his family. By 2010, his son CJ and daughter-in-law Jean were fully ensconced in the business and had expanded the farm to include vegetable production. As the second generation planned to take over the business, they decided the best option was to build a farm store, sell through a CSA and expand their pig operations to start selling meat.

Even though the Knights had been farming for decades, their growth strategy was a departure from what they already knew. They went through each revenue stream to consider what they could sell.

For the CSA, CJ and Jean estimated how many of each type of share they would sell, and at what price.

- 6 full shares at $550/each = $3,300
- 16 half shares at $375/each = $6,000
- 8 single shares at $250/each = $2,000

For a total of $11,300 CSA revenue.

For the piglets, they estimated how many they wanted to sell in a year and for what price.

- 180 feeder pigs at $100 each = $18,000

For the farm store, they estimated how many customers would shop each day, how much each customer would spend and how many days per year they would be open.

They estimated that 25 people would shop each day and would spend, on average, $30. If the stand is open 140 days a year, then:

$$25 \text{ customers} \times 30 \text{ dollars per customer} \times 140 \text{ days} = \$105,000$$

Finally, they decided how many pigs they would raise, and then estimated the average yield and average price per pound.

They would raise 20 pigs; with a finished weight of 200 pounds each. At an average sales price of $6.50 per pound, they would yield $26,000 in revenue.

With each of these sales channels broken out to its base components, the Knights could project a revenue estimate.

Revenue Assumptions — Bottom Up

CSAs	Full	Half	Single	
Number of Shares	6	16	8	
Price per share	$550	$375	$250	
Revenue	$3,300	$6,000	$2,000	$11,300

Piglets		
piglets sold	180	
Sale Price Per Piglet	$100	
Revenue		$18,000

Farm Stand		
Customers Per Day	$25	
Average Purchase	$30	
Days Open Per Year	140	
Revenue		$105,000

Finished Pigs		
pigs raised per year	20	
pounds finished weight	200	
per pound revenue	$6.50	
Revenue		$26,000

Gut Check

For Kristen, she had enough experience working on other people's farms as well as her own, to feel comfortable with her sales projections. The farmers market sales target was ambitious, but she knew that the Copley Square market attracted thousands of customers each week, and with a well laid out display, she'd be able to draw customers.

For a new or beginning farmer, with limited sales experience, assessing the likelihood of achieving these numbers can be challenging, and the best you can do is a gut check. After you've determined how much you will earn from each revenue stream and how the numbers will break down, sit with it. How does it *feel*? Do the numbers feel doable or totally audacious?

Tip: If the sales projections seem optimistic, even if doable, scale them back. There's no shame in conservative numbers. Better to create a business model that works in less than ideal circumstances than one that relies on achieving aggressive goals to succeed.

Estimate Expenses

If you have historical numbers: Go back to your historical financials that show each expense as a percentage of sales. Review each line item.

For variable expenses, look at the history to see the pattern or trends. While numbers can vary from year to year, you may have insight as to why the numbers are not consistent. For Kristen, her seed expenses were 3.0% 1.5% and 2.4% of revenue (in years 2010, 2011 and 2012, respectively). To see the expense go down as a percent of revenue might seem odd on the surface; but I asked her about this, and she explained that she over-purchased in her first year and used up seed inventory in the second year. A reasonable number for her may be 2% – 2.5% of revenue. Similarly, the costs of sales (COGS) decreased as a percentage of sales, from 25.6% to 14.6%. This resulted from selling more of her own product rather than reselling others. Kristen decided that she would continue to resell other farmers' products alongside her own; more than 2012, but less than 2011.

- For each variable expense, estimate what percentage of sales you expect it to be for coming years.
- For fixed numbers, such as accounting services or insurance, estimate your expected expenses for coming years based on what you spent in previous years.

The following table shows Kristen's expense projections.

	Income Statement			Percentage of Sales			Projections	
	2010	2011	2012	2010	2011	2012	Variable	Fixed
Revenue								
Crop Sales	33,862	49,298	148,662	52.4%	55.9%	100.0%		
Resold Crops	29,436	38,886	—	45.5%	44.1%	0.0%		
Holiday Sale	1,330	—	—	2.1%	0.0%	0.0%		
Total Revenue	64,628	88,184	148,662	100.0%	100.0%	100.0%		
Cost of Sales	16,566	24,026	21,768	25.6%	27.2%	14.6%	20.0%	
Gross Profit	48,062	64,158	126,894	0.0%	0.0%	0.0%		
Direct Operating Expenses								
Booth Fees	1,380	921	2,155	2.1%	1.0%	1.4%		2,500
Chemicals	632	86	—	1.0%	0.1%	0.0%	0.0%	
Equipment	731	—	1,983	1.1%	0.0%	1.3%	2.0%	
Fertilizer and Lime	634	1,114	1,045	1.0%	1.3%	0.7%	1.0%	
Mulch	—	—	5,050	0.0%	0.0%	3.4%	3.5%	
Pest Control	—	—	1,381	0.0%	0.0%	0.9%		
Rent or Lease: Vehicle, Machinery and Equipment	—	166	—	0.0%	0.2%	0.0%		
Seeds and Plants	1,936	1,356	3,594	3.0%	1.5%	2.4%	2.0%	
Soil Tests	—	—	—	0.0%	0.0%	0.0%		500
Supplies	2,959	4,620	8,889	4.6%	5.2%	6.0%	5.0%	
Total Direct Operating	8,272	8,263	24,097	12.8%	9.4%	16.2%		
Payroll								
Labor Hire	—	—	42,283	0.0%	0.0%	28.4%	27.0%	
Taxes: Payroll	—	—	—	0.0%	0.0%	0.0%	11.0%	
Worker's Comp + Disability	—	—	1,871	0.0%	0.0%	1.3%		2,000
Total Payroll	—	—	44,154	0.0%	0.0%	29.7%		
General and Administrative								
Accounting Services	—	—	265	0.0%	0.0%	0.2%		300
Advertising	306	69	1,081	0.5%	0.1%	0.7%	0.5%	
Bank Fees	—	—	440	0.0%	0.0%	0.3%		144
Insurance (excluding health)	360	1,276	749	0.6%	1.4%	0.5%		800
Continuing Education	744	29	30	1.2%	0.0%	0.0%		100
Meals and entertainment	60	34	164	0.1%	0.0%	0.1%	0.1%	
Office Supplies	1,244	222	772	1.9%	0.3%	0.5%	0.5%	
Other Expense	—	1,848	—	0.0%	2.1%	0.0%		
Permits and Licenses	266	110	10	0.4%	0.1%	0.0%		50
Professional Fees	—	125	450	0.0%	0.1%	0.3%		500
Shipping	156	32	—	0.2%	0.0%	0.0%		
Storage and Warehousing	2,659	—	—	4.1%	0.0%	0.0%		
Total General and Administrative	5,795	3,745	3,961	9.0%	4.2%	2.7%		
Repairs and Maintenance								
Car and Truck	3,216	5,390	3,847	5.0%	6.1%	2.6%	2.6%	
Gasoline	—	5,238	4,837	0.0%	5.9%	3.3%	4.0%	
Repairs and Maintenance	1,342	1,806	292	2.1%	2.0%	0.2%	0.5%	
Tools	731	306	1,347	1.1%	0.3%	0.9%	0.8%	
Total Repairs and Maintenance	5,289	12,740	10,323	8.2%	14.4%	6.9%		
Occupancy								
Rent or Lease: Other	—	1,226	3,325	0.0%	1.4%	2.2%		3,325
Utilities	528	—	3,000	0.8%	0.0%	2.0%		3,000
Total Occupancy	528	1,226	6,325	0.8%	1.4%	4.3%		

When creating the income statement projections, the variable expenses (including cost of goods sold) will be calculated as a percentage of her revenue. The fixed expenses are listed as annual expenses and will be just as they are listed here.

A note about Cost of Goods Sold (COGS): Not all farms will have COGS. A farm that sells only its own product and no value-added products won't have any. If you sell another farm's product, then estimate the cost of the product relative to the price you sell it for. If you buy apples for $.60/pound and sell them for $1.00/pound, then your COGS is 60% ($.60/$1.00). Not all products will have the same markup: apples may have a 60% COGS and corn 50%. For the purposes of the projections, I recommend using an average for COGS, such as 55% in this case.

Further, only a portion of what you sell will be another farm's products. As such, you only apply the COGS to the portion of product that is resold. This can be simplified by estimating the portion of revenue you expect from reselling products. If you expect to earn 40% of your revenue from reselling product, and your COGS for that 40% is 55%; then the COGS as a variable expense will be:

$$40\% \times 55\% = 22\%\text{[3]}$$

The variable expense applied to your entire sales estimate will be 22%.

If you are launching a new venture: Review the list of expenses on the sample income statement (Appendix 5). We will address labor separately below.

- Which of these expenses do you anticipate having?
- Are there expenses not suggested that you think you'll have?
- For each expense, estimate what you'll spend each month. Where possible, research and get quotes: how much does feed cost, and how much do you think you need? What kind of insurance will you need and how much will it cost? Call a payroll company to understand payroll expenses. How much do black plastic, seed trays and soil amendments cost? How will you package your product, and what does it cost? It's tempting to make up numbers, what you *think* will happen, but there's no substitute for actual numbers and real research.
- Consider how often you will pay for these expenses. Some things, like seed purchases or liability insurance, may be paid once or twice a year. Others you pay once a month, like your rent, utilities or phone bill.
- Create a grid. Down the left side, write down all your expense categories. Across the top, create a column for each month. Then fill in the grid with each of the expenses you anticipate throughout the course of one calendar year. (See table on the next page.)

Total each row. For example, how much in total will you spend on your telephone bill? On animal feed and bedding?

It may look something like the following table:

	Jan	Feb	Mar	Apr	May	Jun	Jul	Aug	Sep	Oct	Nov	Dec	2013
Expenses													
Direct Operating													
Animal Feed			895	1,330			1,330						3,555
Car and Truck Expenses			100	3,000		592				500			4,192
Depreciation Expense													—
Fertilizer/Compost	600	1,200	791				158				150		2,899
Gasoline/Oil/Fuel/Propane	253	253	253	253	253	253	253	253	253	253	253	253	3,036
Greenhouse Expenses													—
Organic Certification		1,080											1,080
Seeds and Seedlings	1,600	700				50			349	200			2,899
Top Soil/Potting Soil		773											773
Cover Crop Seed	255	470								241			966
Ag. Supplies	982	500	300	418	182	200	400	200		200			3,382
Packaging								193					193
Other Direct Operating					473			300					773
Total Direct Operating	3,690	4,976	2,339	5,001	908	1,095	2,334	753	602	1,394	403	253	23,748
Payroll and Benefits													
Salary Expense							5,000	5,000	5,000	5,000	5,000	5,000	30,000
Hourly Labor Expense				300	1,080	1,880	1,880	1,800	1,080	632	400		9,052
Payroll Taxes	—	—	—	23	82	143	523	517	462	428	410	380	2,968
Health Insurance	279	279	279	279	279	279	279	279	279	279	279	279	3,348
Employee Meals	20									20	17	0	57
Other Payroll & Benefits													—
Total Payroll & Benefits	299	279	279	602	1,441	2,302	7,682	7,596	6,821	6,359	6,106	5,659	45,425
Occupancy													
Rent				500									500
Utilities: Gas and Electric						1,000							1,000
Other Occupancy Expense													—
Total Occupancy Expense	—	—	—	500	—	1,000	—	—	—	—	—	—	1,500
Repairs and Maintenance													
Equipment	116	116	116	116	116	116	116	116	116	116	116	116	1,396
Buildings					125						125		250
Coolers/Refrigeration				475									475
Greenhouse(s)													—
Other Repairs & Maintenance			125							125			250
Total Repairs & Maintenance	116	116	241	591	241	116	116	116	116	241	241	116	2,371
General and Administrative													
Bank Charges													—
Conferences/Subscriptions	50	60									475		585
Insurance - Auto							645						645
Insurance - Liability				400	399								799
Office Equipment and Supplies		60				57							117
Postage		30					34						64
Telephone	79	79	79	79	79	79	79	79	79	79	79	79	945
Web-hosting/Internet	51	51	51	51	51	51	51	51	51	51	51	51	611
Other	40	40	40	40	40	40	40	40	40	40	40	40	485
Total General & Administrative	220	320	170	570	569	227	849	170	170	170	645	170	4,251
Advertising and Promotion													
Advertising Expense						150		150					300
Farmers Market Stall Fees					500	499							999
Total Advertising Expense	—	—	—	—	500	649	—	150	—	—	—	—	1,299

Estimate Labor Cost

One of your biggest expenses will be payroll. If you set a budget for $10,000, how many employees can you have? Is that enough to support your sales goals? Will you pay payroll taxes and worker's comp insurance? As your business grows, your payroll expense will also grow.

A sample schedule provides a framework for estimating labor expense.

- List the different positions you need filled—harvesters, CSA managers, farmers market crew, drivers, etc. You don't necessarily need to know who's going to be working, just that someone will be working.
- Set the hourly wage for each position.
- How many people do you need working each shift, each day?
- How many hours will they work in a day? How many days per week?

As you create this sample schedule, consider your harvest schedule and your sales schedule (such as when you will attend farmers markets or open your farm store).

Staffing Needs	Shift Length (hrs)	Hourly Wage	Number of employees per shift							Total Hours	Weekly Expense	Monthly Expense	Seasonal Expense
			Mon	Tue	Wed	Thur	Fri	Sat	Sun				
6:30–3:00													
Field Crew	8.5	$8.00	3	3	3	3	3	1	1	145	$1,156	$4,971	$39,766
CSA Manager	8.5	$8.00	—	—	1	1	1	—	—	26	$204	$877	$7,018
8:00 am–3 pm													
Farm Store	8.0	$8.00	1	1	1	1	1	2	2	72	$576	$2,477	$19,814
5:00 am–2:00 pm													
Driver	9.0	$14.00	1	1	—	1	1	1	—	45	$630	$2,709	$21,672
2:30 pm–8 pm													
Farm Store	5.5	$12.00	1	1	1	1	2	2	2	55	$660	$2,838	$22,704
Total										342	$3,226	$13,872	$110,974

You can find this labor template at juliashanks.com/TheFarmersOfficeTemplates/

Once you determine the total shifts worked per week, the duration of each shift and the hourly pay, you can then calculate the weekly payroll expense. Calculate the annual expense by multiplying the number of weeks your farm is in operation during the year by the weekly labor cost. In this example, the total hourly staff pay is $27,421.

But, wait! There's more! In addition to the wages you pay to your employees, you will also pay federal and state taxes, social security, worker's comp insurance and unemployment insurance on behalf of your employees. These contributions are calculated as a percentage of total salaries and wages and can vary from 11% to 18% in additional expense.[4]

The six categories are:
- FUTA (Federal Income Tax) .6%
- Medicare 1.45%
- Social Security (FICA) 6.2%
- SUTA (State Income Tax)—for state tax rates, visit: payroll-taxes.com/state-tax
- Worker's Comp Insurance
- Unemployment Insurance

For simplicity sake when creating financial projections, add 11% to the salaries for payroll withholdings and insurance. For actual day-to-day operations, I strongly encourage you to seek a payroll-service provider to ensure that you are properly withholding payroll taxes and submitting them to the appropriate agencies.

As your business grows, presumably so will your payroll. As you raise more chickens, milk more cows, harvest more produce, you will need a larger crew. The payroll calculation above can be used as the baseline for future years by calculating the payroll number as a percentage of sales.

For example, if you anticipate wages to be $27,000 in your first year, and your revenue to be $130,000 then wages as a percentage of sales is 20.7%. In subsequent years of your projections, you can use 20.7% to estimate your payroll expense.

3. Calculate Your Projected Operating Profit

Start with the Income Statement Projections template (juliashanks.com/TheFarm ersOfficeTemplates/): enter the fixed and variable costs into the appropriate columns (see instructions in the spreadsheet) as well as your revenue assumptions. With revenue and expense assumptions laid out, you can calculate down to the operating profit. If you don't have access to the template, then you can create the projections manually.

The projections are created not just for one year, but for 3 to 5 years. The assumptions you created above may be focused primarily on the first year.

To create years 2 to 5, look back at year 1:
- For revenue, consider your sales goals. How will your sales increase? You may want to repeat the process used for year 1 to calculate sales projections.
- For each expense considered variable as it relates to revenue, calculate its percentage of sales. Will the cost remain constant as a percentage of sales? Do you expect to gain efficiencies in future years?
  - Let's say you project packaging expense to be $2,900 in year 1 with $130,000 in revenue. The percentage is 2.2% (2,900/130,000). For year 2, you can project the packaging expense to be 2.2% of the projected revenue.
  - Perhaps in year 2, your business volume increases sufficiently that you can get volume discount pricing for your boxes. You may reduce the packaging expense to 1.7%.

Entrepreneurial Thinking

A few other expenses will be factored into the income statement after calculating the operating income (interest expenses, taxes, depreciation and amortization), but this is a good place to stop and evaluate the venture and projections. At its core, does your business model make sense and cents? Some questions to ask:

- Can you support yourself from the profits and/or pay yourself a salary?
- Are the profits sufficient to grow the business? Should you decide to build a new chicken coop or milking parlor, can you afford to save money?
- If the business experiences a hiccup (the tractor needs an unexpected repair), will you have enough breathing room financially?

As an entrepreneur, consider whether the business generates enough profit to satisfy your financial needs: grow your business, pay your debt and withstand the unexpected expense. Also consider that you may crave the occasional luxury (new Carhartts or a dinner out) and the capacity to enjoy a vacation every once in a while. Sure, you love what you do but that elusive "work-life balance" is important too.

Further, the operating profit also begins to give an indication of how much debt financing the business can afford. The monthly loan payments certainly cannot exceed the operating profits, or you will run into cash flow problems; better still, the debt service should still leave breathing room for the above.

If the business is not as profitable as you think it should be at this stage of the projections, then take a break and revisit your assumptions. Profits will only shrink further when you factor in debt service and taxes. In addition, things never go as expected—though you carefully planned out your assumptions, something will be off. You won't get into a farmers market, your first planting of tomatoes will get flooded by a heavy rain, or a predator will get into the chicken coop. The projections need enough wiggle room to absorb the inevitable setbacks.

After operating profit ("below the line") come other income and other expenses. Other income includes grants,[5] off-farm income and profits from the sale of equipment or other assets. List whatever other income you expect over the course of the next five years.

We'll take a break from the Profit and Loss statement now.

4. Investments in Your Business

The P&L can't be finished until the depreciation, interest and tax expenses are estimated. To calculate the depreciation, you need to list the capital improvements and purchases you will make. To calculate the interest expenses, you will estimate the loans you will take to finance the purchases and improvements. Together, these become your **Sources and Uses of Funds**.

Uses of Funds

If you're building a farm store, you may need funds for labor and materials, installing a septic system, grading a driveway and parking area and so on. Or perhaps, all you need to purchase is a tractor and a disc harrow. This list will become your **Uses of Funds.**

This table shows Kristen's Uses of Funds.

	2013	2014	2015
Mower	2,500	—	—
Box Truck	15,000	—	—
Greenhouse	8,000	—	—
Walk-In Cooler	3,500	—	—
Implements	3,000	1,000	—
Tractor	—	—	750
Tractor-Trailer	—	1,500	—

Knight Farm, which was building a farm store, had a more elaborate list.

	2011	2012	2013
Farm Store			
Supplies/Materials	$23,500		
Labor	$10,000		
Plumbing/Electrical	$15,000		
Interior Barn Board	$ —		
Shelving and Displays	$1,500		
Refrigeration			
2x Refrigerators	$3,000		
2x Freezers	$1,500		
Septic System			
Engineer Drawings	$5,000		
Installation	$25,000		
Pig Housing			
Supplies			$30,000
Labor			
Gravel Road			
Supplies	$3,000		
Labor			
Greenhouse			$10,000
Total	$87,500	$ —	$40,000

Use of Funds and Depreciation Expense

With the list of planned improvements, determine which projects or purchases to note on the balance sheet; that is, which are assets (versus expenses which appear on the income statement) and then calculate their annual depreciation expense. As a general rule, purchases that have a useful life of more than a year, such as a tractor, greenhouse or large equipment (with the exception of land) depreciate. Labor expense (as Knight Farm has) to construct an asset (the farm store) can be folded into the value of the farm store (in which case it is *capitalized*), or written as an expense on the income statement projections. For Knight Farm, all items on the list will be depreciated with the exception of the labor. For Kristen, all purchases will depreciate.

For the projections, it is not too important to get the depreciation calculation perfect as it is a non-cash expense, and it will not impact your cash position should you make a mistake in the calculation. Your tax estimates will be off, but not too much. The depreciation expense acts as a reminder that you should be saving that amount to replace the assets. If you can afford to save that amount of money, you won't need to borrow money when it comes time to replace the tractor, or whatever asset needs replacing. When filing your taxes, you'll want to be more precise; and I recommend that you talk with your accountant.

To estimate the annual deprecation:

1. For each asset, determine its useful life. You can use the asset class table provided by the IRS,[6] talk with your accountant or make an estimate.
2. Take the purchase price of the asset and divide by the number of years that asset will last.

For example, a greenhouse (according to the IRS) has a 10-year life. Since it costs Kristen $10,000, she will depreciate $1,000 a year for 10 years. Use the **depreciation template** to estimate total annual depreciation (juliashanks.com/TheFarmers OfficeTemplates/).

Sources of Funds

This list of purchases and improvements (uses of funds) also informs your options on financing. Do you need to borrow money, get a grant, or can you finance the purchases with earnings from the business? Write out where you think you will get financing. This becomes your **Sources of Funds**.

Interest Expense

If you borrow money, you will repay the money with interest. Depending on the stage of your planning, you may or may not know for what type of loan you qualify. At the very least, call a few banks to get interest rates and loan terms. Use the **Debt**

Service template to calculate the monthly payments.[7] (juliashanks.com/TheFarmers OfficeTemplates/)

As you pay back the loan, each installment will include a portion of the principal and interest. In proper accounting, the interest expense is noted on the income statement, and the principal repayment is noted on the Balance Sheet. With each payment, the split between interest and principal changes; it can be difficult to keep track of what goes where. The **Debt Service** template will help you keep track of this.

Tax Expense

With depreciation and interest expense estimated, you can also estimate your income/business tax. From the operating profit calculated in step 3 above, subtract depreciation and interest expense. This subtotal (earnings before taxes) is used to estimate tax payments.

While you may be estimating profits on a monthly basis, taxes are estimated based on the annual total. Let's say that from January to April you estimate a net loss of $5,000. From May to December, you estimate a net profit of $25,000. If you calculate taxes each month, then you'd have zero tax payments for the first 4 months, and estimate taxes for the last 8 months based on $25,000 in earnings. If you estimate taxes on the total year, taxes will be based on $20,000 in earnings.

Estimate your annual taxes based on your local tax rate. If you are unsure about this number, consult with your tax accountant. While talking with your tax accountant, ask whether you should make estimated quarterly tax payments.

Whether the business pays the taxes, or you personally pay the taxes on the profit, they will be paid and your cash flow from operations will be impacted.

5. Calculate Net Income

With all the numbers necessary, you can complete the income statement projections. From the operating profit, you will add other income and subtract other expenses (depreciation and interest). From the subtotal (EBT, Earnings Before Taxes), subtract taxes. This brings you to the net income.

Kristen's income statement projections looked like this. For year 1, she estimated revenue and expenses by month, and years 2 and 3 she estimated an annual summary.

To summarize what we did:

Revenue – COGS = Gross Profit
Gross Profit – Operating Expenses = Operating Profit
Operating Profit +/- Other Income and Expenses = Earnings Before Taxes
Earnings Before Taxes – Taxes = Net Income

	Jan	Feb	Mar	Apr	May	Jun	Jul	Aug	Sep	Oct	Nov	Dec	Total 2013	2014	2015
Revenue															
CSA	5,000	5,000	5,000	5,000									20,000	21,400	23,112
Farmers Markets						20,000	23,000	32,000	29,000	25,000	13,000	7,000	149,000	150,870	158,414
Wholesale				500	500	1,000	1,500	3,000	2,000	2,000	1,000	500	12,000	14,400	16,560
Seedlings				500	2,000								2,500	2,625	2,704
XMAS												2,500	2,500	2,750	2,888
Total (cash) Revenue	5,000	5,000	5,000	6,000	2,500	21,000	24,500	35,000	31,000	27,000	14,000	10,000	186,000	192,045	203,677
Cost of Goods Sold						3,000	4,500	5,500	5,500	5,000	1,000	1,000	25,500	26,886	28,515
Direct Operating Expenses															
Booth Fees		300	300	1,200		500		500		200	500		3,500	3,675	3,859
Fertilizer and Lime			1,500				500						2,000	2,112	2,240
Mulch				2,500					2,000				4,500	4,801	5,092
Pest Control				300	500	250	250	250					1,550	1,728	1,833
Propane			150	150	150								450	468	487
Seeds and Plants	2,500		200	800		200		200		200			4,100	4,417	4,685
Soil Tests			100										100	200	200
Supplies (including soil)		1,200	1,000	3,000	2,250	1,000	500	500	1,000	250	250	250	11,200	12,080	12,811
Total Direct Operating	2,500	1,500	3,250	7,950	2,900	1,950	1,250	1,450	3,000	650	750	250	27,400	29,482	31,207
Payroll															
Labor Hire			1,600	3,000	5,200	10,600	10,600	10,600	7,600	7,600	4,400	1,600	62,800	62,800	62,800
Management Salary	1,000	1,500	2,000	2,000	2,000	2,000	3,000	3,000	3,000	4,000	4,000	4,000	31,500	33,075	34,729
Taxes: Payroll	76	114	274	380	547	958	1,034	1,034	806	882	638	426	7,167	7,287	7,412
Worker's Comp—Disability	82	82	82	82	2,000	82	82	82	82	82	82	82	2,902	3,000	3,000
Total Payroll	1,158	1,696	3,956	5,462	9,747	13,640	14,716	14,716	11,488	12,564	9,120	6,108	104,369	106,162	107,941
General and Administrative															
Accounting Services			600										600	600	600
Advertising	50	50	50	50	150	50	50	50	50	50	50	50	700	735	772
Bank Fees	750												750	145	145
Insurance (vehicles and liability)				1,000	650				1,600				3,250	3,348	3,448
Continuing Education		100	100										200	200	200
Meals and Entertainment													—	192	204
Office Supplies		150		150		150		150		150		50	800	840	882
Permits and Licenses													—	—	—
Professional Fees			50	350	50	50	50	50	50	50	50	50	800	840	882
Subscriptions		100											100	100	100
PO Box + Shipping			144										144	144	144
Total General and Administrative	800	400	944	1,550	850	250	100	250	1,700	250	100	150	7,344	7,143	7,377
Repairs and Maintenance															
Car and Truck	400		200			200		200		200		200	1,400	1,536	1,629
Gasoline	350	350	350	350	450	600	625	650	650	650	500	500	6,025	6,722	7,129
Motor vehicle					1,000								1,000	1,075	1,141
Repairs and Maintenance			100		100		100			100			400	960	1,018
Tools		500		200		200		200		200			1,300	1,536	1,629
Total Repairs and Maintenance	750	850	650	550	1,550	1,000	725	1,050	650	1,150	500	700	10,125	11,830	12,546
Occupancy															
Rent or Lease: Other				300	300	300	300						1,200	1,200	1,200
Utilities		300										300	600	612	624
Total Occupancy	—	300	—	300	300	300	300	—	—	—	—	300	1,800	1,812	1,824
Total Operating Expenses	5,208	4,746	8,800	15,812	15,347	17,140	17,091	17,466	16,838	14,614	10,470	7,508	151,038	156,428	160,895
Operating Income	(208)	254	(3,800)	(9,812)	(12,847)	860	2,909	12,034	8,662	7,386	2,530	1,492	9,462	8,730	14,267
Depreciation	518	518	518	518	518	518	518	518	518	518	518	518	6,214	6,214	6,214
Interest Expense	—	—	—	156	154	152	150	148	145	143	141	139	1,328	1,484	1,133
Taxes															
Net Income	(726)	(264)	(4,317)	(10,486)	(13,519)	191	2,242	11,369	7,999	6,725	1,871	836	1,920	1,032	6,919

Entrepreneurial Thinking

While it is normal to have net losses in the first year or two, the business should turn a profit by year 3. If the business does not project profit within a time frame that feels comfortable to you, then revisit your assumptions.

It is possible to have a positive operating profit and negative net income. This results from interest payments that exceed what the business can afford and/or depreciation expense. As you create the balance sheet and cash flow projections, pay close attention to the cash balance to make sure it stays positive.

Kristen shows a positive net income, but it may not be great enough to service her debt and save for future investments. She will want to pay attention to how her cash balance changes and what she can afford to set aside to grow the business.

Further, the net income does not look as strong as it did in previous years. For Kristen's projections, she included her own salary in the operating expenses, which she did not in previous years. With her salary factored in, a positive net income shows that she can afford to pay herself.

6. Create the Balance Sheet

Projections for the Balance Sheet start from the historical (beginning) balance sheet statement, and then layer in new purchases, financing and ongoing profits. (See also, the Balance Sheet template at juliashanks.com/TheFarmersOfficeTemplates/)[8]

If you have an existing business, you'll start with a current balance sheet to then create projections of what your balance sheet will look like as a result of expanding and growing operations. If you are starting a new venture, you will create a balance sheet for the day of inception; that is, the day you swing open the doors of your business.

To create a starting balance sheet for an existing business:

- If you're using QuickBooks, run the balance sheet report (**Reports > Company and Financials > Balance Sheet Standard**).
- If you don't have QuickBooks, simply create the balance sheet by listing assets and liabilities. Owner's equity is calculated by subtracting liabilities from assets.

Owner's Equity = Assets – Liabilities

For new businesses, without any historical balance sheet, you will create a "beginning balance sheet" by estimating what you anticipate you will have the day you open your business. Some examples are:

- equipment you will you purchase and the purchase price
- debt you will take on to launch the business
- money you will have in the bank after the loans are received and the purchases are made

The Beginning Balance Sheet

You can also use the Balance Sheet template. For a refresher on the Balance Sheet, review Chapter 2: Building the Foundation: The Financial Statements and Basic Accounting.

Whether you create a beginning balance sheet for a new business or for a current business, the process is the same. For a new business, you will want to project what you expect to have on the day you start your business (after you have made the initial purchase of equipment and received financing).

1. Start with Your Assets

As you list assets with their values, disregard any debt associated with them. The debt will be noted in the liabilities section. For example, if you purchased a truck for $15,000 and took out a loan for $10,000 to pay for it (using $5,000 of your own money), the truck should be noted on the balance sheet with a value of $15,000. The loan will be listed in the liabilities section as $10,000.

   a. List bank balances for your business accounts (not personal), including checking and savings accounts. For a new business, list the amount of cash you expect to have in your bank account *after* purchasing equipment, vehicles, structures and land.

   b. Note how much customers owe you (Accounts Receivable). For a new business, you should not have any A/R.

   c. List equipment that is expected to last another year as well as their values. Equipment can include tractors, tillers, seeding and cultivation equipment, refrigerators and coolers. Be sure to list their current value regardless of whether you owe money or not.

   d. List trucks and other vehicles, as well as their value. Be sure to list their current value regardless of whether you owe money or not.

   e. List structures and buildings such as greenhouses and barns that you built/purchased separate from the land you use/own. Be sure to list their current value regardless of whether you owe money or not.

   f. List land and buildings and their purchase price.

As discussed earlier, most assets[9] lose their value (depreciate) over time. The truck you purchased three years ago for $25,000 may only be worth $10,000 today. With proper accounting, you would list the original purchase price along with the accumulated depreciation. If you have not tracked depreciation before now, then it is not worth the effort to go back and determine the accumulated depreciation for all your assets. Going forward, you can add it in. With the beginning balance sheet, I suggest you start with the current value of your assets.

Assets can be grouped into the following categories, with subtotals for each section:

- Cash and Current Assets
- Equipment
- Vehicles
- Fixed Assets (Land, Buildings and Real Estate)

After listing all your assets, add up the values and note that total at the bottom of the Assets section.

2. List Your Liabilities

Liabilities should be listed in the order for which they are due. Liabilities may also include any debt or loans associated with above-listed assets.

a. List total amount due to vendors (Accounts Payable). If you have a new business, you may have terms with your vendors (hardware store, seed supplier, etc.) to pay later.
b. List any credit card debt that your business has.
c. List any short term loans such as lines of credit or operating loans. Only list the money you owe, not the amount available to you.
d. List medium term debt, such as car loans.
e. List long-term debt such as mortgages. Only list mortgages associated with the business.

Liabilities can be grouped with subtotals:
- Current Liabilities (A/P, credit card debt and lines of credit)
- Long-term Liabilities

3. Calculate Owner's Equity

The amount of money you invested in your business may not represent the actual equity you have. Money spent on a website, that first order of seeds and animal feed; these are real expenses of the business, but are not represented in the assets and liabilities section of the balance sheet. They diminish the equity in the business.

Owner's equity as a formula equals assets minus liabilities. Owner's equity as a concept is the money you put into the business plus (or minus) the profits (or losses). If you put $10,000 into your business, but have $2,000 in start-up expenses, your equity is $8,000.

Option 1 (*Easy*)

For the beginning balance sheet, you can calculate owner's equity simply by using this formula:

$$\text{Owner's Equity} = \text{Total Assets} - \text{Total Liabilities}$$

Option 2 (Less Easy)

You may prefer to track the amount of money you invested in your business. In addition to noting your initial (and subsequent investments), you will note the **retained earnings**. In the start-up phase of a new venture, the retained earnings amount is usually negative—it is the net losses in the business. When the business is starting up, it has expenses but not revenues; hence the loss and the negative retained earnings.

For a farmer who invested $10,000 into her business and had $7,343 in start-up expense, the owner's equity portion of the balance sheet would look like this:

Owner's Equity	
Retained Earnings	(7,343)
Owner's Equity	10,000
Total Equity	2,657

No matter how you calculate **owner's equity**, it should still equal **Total Assets – Total Liabilities**.

Kristen's beginning balance sheet looked like this. She had been tracking the depreciation for her equipment, so she listed the accumulated depreciation along with the purchase price of the equipment.

	Beginning 12/31/2012
Assets	
Current Assets	
Cash	5,000
Inventory	5,500
Vehicles	
Ford F150	5,500
Toyota Rav 4	17,000
Equipment	
BCS Tractor with Implements	7,000
Hand Tools	3,000
Greenhouses	5,500
Fencing	2,000
Walk-In Cooler	3,000
Farmers Market Supplies	6,000
Accumulated Depreciation	(4,000)
Total Assets	55,500
Liabilities	
Debt	10,750
Toyota Rav 4 Loan	8,000
Total Liabilities	18,750
Equity	
Owner's Equity	36,750
Total Equity	36,750
Total L + OE	55,500

Balance Sheet Projections

Create the balance sheet projections in the same interval as the income statement projections. If you project income for each month, then your balance sheet projections should be by month.

The balance sheet will be an ongoing tally of your assets with an accounting of how you purchased the assets: did you borrow money, reinvest earnings from the business, get a grant or get investors? It communicates your projected financial position.

As part of creating the income statement (in step 4), you created a uses of funds list. Now you need to align the uses with a source. How are you going to finance all those purchases? Will you get loans, grants or use cash generated through your business operations? This informs the liabilities and equity sections of your balance sheet.

	Uses of Funds				Sources of Funds				
	2013	2014	2015	Total	Operating	Grants	Loan 1	Loan 2	Total
Mower	2,500	—	—				2,500		
Box Truck	15,000	—	—			9,250	5,750		
Greenhouse	8,000	—	—					8,000	
Walk-In Cooler	3,500	—	—		—		1,500	2,000	
Implements	3,000	1,000	—		1,000	3,000	—		
Tractor	—	—	750		—		750		
Tractor-Trailer	—	1,500	—		1,500		—		
Total	32,000	2,500	750	**35,250**	2,500	12,250	10,500	10,000	**35,250**

The **uses of funds** details the investments in assets; and the **sources of funds** notes whether the entrepreneur will incur a liability (from loans) or use the equity of the business (operating profits or grants).

Lay out the balance sheet projections in an Excel spreadsheet with the beginning balance sheet in the first column. Then create subsequent columns for each month of projections. In the month where you think you will make a purchase, increase the asset account by the amount of the purchase.

For example, if you purchase the greenhouse in January, and already have a greenhouse valued at $5,500, then the value of your greenhouses increases by $8,000 to $13,500.

	12/31/2012	1/31/2013
Assets		
Current Assets		
Cash	5,000	13,703
Inventory	5,500	6,200
Vehicles		
Ford F150	5,500	5,500
Toyota Rav 4	17,000	17,000
Box Truck	—	15,000
Accumulated Depreciation	(3,000)	(3,000)
Equipment		
BCS Tractor with Implements	7,000	7,000
Hand Tools	3,000	3,000
Greenhouses	5,500	13,500

Remember, the balance sheet reflects the total value of your assets, so you want to include what you already had plus what you purchase. For each month, note the increased value of your assets as you project when you will purchase them.

Similarly, as you take out loans, note the *current* value of the loan. In the example below, Kristen took out two new loans to finance her purchases. In addition, she paid down her outstanding debts.

Liabilities		
Debt	$10,750	$10,650
Toyota Rav 4—Loan	$8,000	$7,764
Loan 1	—	$15,000
Loan 2	—	$10,000
Total Liabilities	$18,750	$43,414

In other words, what would the balance sheet look like if you were to create it for each month going forward?

For balance sheet projections, the owner's equity is calculated by adding or subtracting the profits (or losses) from the previous month's retained earnings. If you have invested more money into the business or taken a draw from the business, note this in the equity section as well.

For example, if the beginning balance of the retained earnings was $11,796, and the business generated profits of $5,134 (this number comes from the **income statement projections** of the same time period), then the ending balance of the Retained Earnings Account would be $16,930 (11,796 + 5,134).

Owner's Equity		
Retained Earnings	11,796	16,930

Note other changes in owner's equity accounts as necessary, such as owner's investment or owner's draw.

For the balance sheet projections, we can simply fill in the different accounts. The one change that we haven't yet noted is the change in cash balances. Since we can project what all the other account balances will be, we can "solve for cash."

Remember the balance sheet equation:

$$\text{Total Assets} = \text{Total Liabilities} + \text{Owner's Equity}$$

This can be written another way:

$$\text{Cash} + \text{All Other Assets} = \text{Total Liabilities} + \text{Owner's Equity}$$

And if we turn this equation around, we can solve for cash:

$$\text{Cash} = (\text{Total Liabilities} + \text{Owner's Equity}) - \text{All Other Assets}$$

Solving for cash provides an estimate for your cash balance.

Getting Fancy—Estimating A/P and A/R

When actually running your business, and not just creating projections of what you think you will do, you will track accounts payable and accounts receivable. In fact, a key component of business and cash flow management is keeping an eye on these two numbers. Incorporating them into your financial projections will give you a better handle on your cash flow and show you where you might run into trouble. (It will also impress your investors that you considered it).

The calculations are trickier than other aspects of the balance sheet, which is why they are often omitted from projections. It is not impossible, though.

- **Accounts Payable:** On *average*, how long do you estimate having to pay your vendors? 7 days, 14 days, 1 month? Some vendors will require payment straightaway (your local hardware store, for example), whereas others may let you pay up to 30 days later. This number is called "Days in A/P." Use the following formula to estimate the *monthly* ending balance in your accounts payable:

Ending A/P Balance =
[(Total Direct Operating Expenses for the month)/30] × Days in A/P

For Kristen, in January, she estimated that her direct operating expenses would be \$3,690. Let's say her average days in A/P was 15; that is, on average, she took 15 days to pay her vendors.

Ending A/P balance for the January balance sheet would be: .

$$= (\$3,690/30) \times 15 = \$1,845$$

That means that for all of Kristen's operating expenses in January, she did not yet pay for \$1,845 worth of them by the end of the month. In other words, for all her expenses that she incurred, she still had \$1,845 in her bank account that had not yet been paid out for bills. From a cash flow perspective, it's great that she has the extra money in her account. But she'll need to be careful that she has the money 15 days hence to pay the bills.

This formula changes slightly if you're calculating the ending A/P balance for an annual balance sheet.

Ending A/P Balance =
[(Total Direct Operating Expenses for the year)/365] × Days in A/P

- **Accounts Receivable:** On *average*, how long do you estimate it will take for your customers to pay you? 7 days, 14 days, 1 month? Some customers will pay straightaway (at the farmers market, for example), whereas others may take up to 30 days later (such as large wholesale accounts). This number is called "Days in A/R." Use the following formula to estimate the monthly ending balance in your accounts receivable:

Ending A/R Balance = [(Total Revenue)/30] × Days in A/R

Kristen anticipated that she would have $5,000 in revenue for January and it would take, on average, three days for her customers to pay.

$$\text{Ending A/R Balance for January} = (\$5,000/30) \times 3 = 500$$

In other words, for all the revenue Kristen *earned*, she would not have $500 of it in her bank account. From a cash-flow planning perspective this could tighten things up. Even though she projected the revenue, it may not be available to pay expenses.

This formula changes slightly if you're calculating the ending A/P balance for an annual balance sheet.

$$\text{Ending A/R Balance} = [(\text{Total Revenue for the year})/365] \times \text{Days in A/R}$$

Entrepreneurial Thinking

As you review your balance sheet projections, watch the cash balance for each month. Does it dip below zero? Mathematically, negative cash is possible but in reality it is not. If your cash balance is negative, then someone is not getting paid and checks will bounce. The result is that you may need an unexpected loan or you may run up credit card debt.

If the cash balance goes negative, and then recovers (that is, you show a positive cash balance in later months), the issue can be the timing of purchases and debt payments. You may need to delay some expenses until you are in a better cash position. If the cash balance dips below zero, but does not return to a positive balance, then there are core issues in your assumptions.

Revisit your work so far. So many assumptions and decisions have been made to get to this point, and many ways in which things went astray that lead to a potential of negative cash. Certainly, this isn't a foregone conclusion—it may turn out that your actual revenues are higher than you project and your expenses are lower than you expect. Ways to troubleshoot a negative cash balance:

- **Is net profit high enough to support all the purchases you want to make and debt you take on?** To that end, are your expenses too high? Are your revenues too low? What changes do you need to make?
- **Did you appropriately capitalize your business?** That is, did you borrow the right amount of money? Did you borrow enough money to carry you through start-up? Are the terms of your loan prohibitively expensive?
- **Does your equipment purchase schedule align with your cash flow?** Do you plan to purchase more equipment than you can afford? Do you need to delay some purchases? What changes can you make to ensure that you project a positive cash balance at all times?

Based on your troubleshooting, revise the assumptions so that you maintain a positive cash balance.

Here's an example of a balance sheet by month:

	Dec-14	Jan	Feb	Mar	Apr	May	Jun	Jul	Aug	Sep	Oct	Nov	Dec
							2015						
Assets													
Checking/Cash	4,335	25,757	23,190	12,478	4,600	4,230	9,159	13,636	8,150	9,795	9,953	8,931	6,331
Accounts Receivable	1,120												
Fixed Assets													
Fencing	2,197	2,197	2,197	2,197	2,197	2,197	2,197	2,197	2,197	2,197	2,197	2,197	2,197
Animal Housing/ Infrastrucutre	3,041	3,041	3,041	3,041	3,041	3,041	3,041	3,041	3,041	3,041	3,041	3,041	3,041
Equipment	2,432	2,432	2,432	2,432	2,432	2,432	2,432	2,432	2,432	2,432	2,432	2,432	2,432
Cold Storage	889	889	889	889	889	889	889	889	889	889	889	889	889
New Purchases													
Hi Tunnel (Layer Winter Housing)		—	—	—	—	—	—	—	—	—	—	—	—
Egg Washer		—	—	2,000	2,000	2,000	2,000	2,000	2,000	2,000	2,000	2,000	2,000
Nest Boxes—Layers		—	1,500	1,500	1,500	1,500	1,500	1,500	1,500	1,500	1,500	1,500	3,000
Coop—Layers		—	3,200	3,200	3,200	3,200	3,200	3,200	3,200	3,200	3,200	3,200	6,400
Buildings and Infrastructure		—	—	1,000	1,000	1,000	1,000	1,000	1,000	1,000	1,000	1,000	1,000
Tractor Equipment		—	—	—	3,500	3,500	3,500	3,500	3,500	3,500	3,500	3,500	3,500
Veg Equipment		—	—	—	—	—	—	—	—	—	—	—	—
Layer Equipment		—	—	1,000	1,000	1,000	1,000	1,000	1,000	1,000	1,000	1,000	1,000
Vehicles	700	700	700	700	8,200	8,200	8,200	8,200	8,200	8,200	8,200	8,200	8,200
Accumulated Depreciation		(496)	(993)	(1,489)	(1,985)	(2,481)	(2,978)	(3,474)	(3,970)	(4,466)	(4,963)	(5,459)	(5,955)
Total Assets	14,714	34,519	36,157	28,948	31,574	30,708	35,140	39,121	33,138	34,288	33,949	32,431	34,035
Liabilities													
Credit Cards	—												
FSA—Micro-loan	6,300	6,300	8,300	8,300	8,300	8,300	8,300	8,300	7,425	7,425	7,425	7,425	7,425
Friendly Loan		9,881	9,762	9,643	9,524	9,405	9,286	9,167	9,048	8,929	8,810	8,690	8,571
FSA New		10,000	10,000	10,000	10,000	10,000	10,000	10,000	9,164	9,164	9,164	9,164	9,164
Total Liabilities	6,300	26,181	28,062	27,943	27,824	27,705	27,586	27,467	25,636	25,517	25,398	25,279	25,160
Equity													
Retained Earnings	8,502	8,426	8,183	1,093	3,838	3,091	7,642	11,742	7,590	8,859	8,639	7,240	8,963
Total Equity	8,414	8,338	8,095	1,005	3,750	3,003	7,554	11,654	7,502	8,771	8,551	7,152	8,875
Total Liabilities + Equity	14,714	34,519	36,157	28,948	31,574	30,708	35,140	39,121	33,138	34,288	33,949	32,431	34,035

7. Create a Statement of Cash Flows

We're in the final stretch of creating financial projections!

The balance sheet tells you if run out of cash or not (among other things); the cash flow statement provides more granularity.

It is divided into three categories:

- Cash Flow from Operations (CFO)
- Cash Flow from Investing (CFI)
- Cash Flow from Financing (CFF)

The most common mistake that entrepreneurs make when creating cash flow projections is that they ignore the investing and financing components—and essentially recreate the income statement in a slightly modified form.

The balance sheet shows *changes* in these accounts from one period to the next. To determine the *cash flow* for a given period of time, you can look at the beginning and ending balance sheets. For cash flow for the month of January, compare the balance sheet from December 31 (of the previous year) with the balance sheet from January 31. The changes in the account balances represent the flows of cash.

							2015						
Assets	Dec-14	Jan	Feb	Mar	Apr	May	Jun	Jul	Aug	Sep	Oct	Nov	Dec
Checking/Cash	4,335	25,757	23,190	12,478	4,600	4,230	9,159	13,636	8,150	9,795	9,953	8,931	6,331
Accounts Receivable	1,120												
Fixed Assets													
Fencing	2,197	2,197	2,197	2,197	2,197	2,197	2,197	2,197	2,197	2,197	2,197	2,197	2,197
Animal Housing/ Infrastrucutre	3,041	3,041	3,041	3,041	3,041	3,041	3,041	3,041	3,041	3,041	3,041	3,041	3,041
Equipment	2,432	2,432	2,432	2,432	2,432	2,432	2,432	2,432	2,432	2,432	2,432	2,432	2,432
Cold Storage	889	889	889	889	889	889	889	889	889	889	889	889	889
New Purchases													
Hi Tunnel (Layer Winter Housing)		—	—	1 —	—	—	—	—	—	—	—	—	—
Egg Washer		—	—	2,000	2,000	2,000	2,000	2,000	2,000	2,000	2,000	2,000	2,000
Nest Boxes—Layers		—	1,500	1,500	1,500	1,500	1,500	1,500	1,500	1,500	1,500	1,500	3,000
Coop—Layers		—	3,200	3,200	3,200	3,200	3,200	3,200	3,200	3,200	3,200	3,200	6,400
Buildings and Infrastructure		—	—	1,000	1,000	1,000	1,000	1,000	1,000	1,000	1,000	1,000	1,000
Tractor Equipment		—	—	—	3,500	3,500	3,500	3,500	3,500	3,500	3,500	3,500	3,500
Veg Equipment		—	—	—	—	—	—	—	—	—	—	—	—
Layer Equipment		—	—	1,000	1,000	1,000	1,000	1,000	1,000	1,000	1,000	1,000	1,000
Vehicles	700	700	700	700	8,200	8,200	8,200	8,200	8,200	8,200	8,200	8,200	8,200
Accumulated Depreciation		(496)	(993)	(1,489)	(1,985)	(2,481)	(2,978)	(3,474)	(3,970)	(4,466)	(4,963)	(5,459)	(5,955)
Total Assets	14,714	34,519	36,157	28,948	31,574	30,708	35,140	39,121	33,138	34,288	33,949	32,431	34,035
Liabilities													
Credit Cards	—							2					
FSA—Micro-loan	6,300	6,300	8,300	8,300	8,300	8,300	8,300	8,300	7,425	7,425	7,425	7,425	7,425
Friendly Loan		9,881	9,762	9,643	9,524	9,405	9,286	9,167	9,048	8,929	8,810	8,690	8,571
FSA New		10,000	10,000	10,000	10,000	10,000	10,000	10,000	9,164	9,164	9,164	9,164	9,164
Total Liabilities	6,300	26,181	28,062	27,943	27,824	27,705	27,586	27,467	25,636	25,517	25,398	25,279	25,160
Equity		3											
Retained Earnings		8,502	8,426	8,183	1,093	3,838	3,091	7,642	11,742	7,590	8,859	8,639	7,240	8,963
Total Equity	8,414	8,338	8,095	1,005	3,750	3,003	7,554	11,654	7,502	8,771	8,551	7,152	8,875
Total Liabilities + Equity	14,714	34,519	36,157	28,948	31,574	30,708	35,140	39,121	33,138	34,288	33,949	32,431	34,035

The months in which a number changes represent a time when there was a cash flow. For example:

1. In March the value of the egg-wash went from zero to $2,000. That indicates a purchase (and cash flow) of $2,000. The cash flow from investing would indicate a $2,000 outflow. The corresponding financing (loan) for that purchase

occurred in February, as witnessed by the increase in the loan balance. That was an *inflow* of cash from financing. You'll notice that the value of the egg-washer does not change for subsequent months. Any lost value of the egg-washer will be noted in accumulated depreciation.

2. In August, the FSA Micro-Loan decreased: it was the month the farmers expected to make a loan payment.
3. Retained earnings fluctuates based on the net income (or loss) of each month

Cash Flow from Operations (CFO)

CFO equals net income plus/minus any non-cash transactions.[10] These transactions include:

- Depreciation: While depreciation appears on the income statement as an expense, it's not an actual cash expense. The actual cash flow associated with the purchase happened months, if not years, ago, and was shown in the cash flow from investing section.
- Accounts Payable: If you track your expenses on an accrual basis, you may have expenses for which you have not yet paid. You may have charged them to your credit card or received terms from your vendors. The expense appears on your income statement but the cash did not flow out of your business (yet!) as a result.
- Accounts Receivable: If you track your revenue on an accrual basis, you may have sales for which you have not yet received cash-money.

Adjusting for A/P and A/R in the Statement of Cash Flows

If you opted against getting fancy with A/P and A/R (i.e., not including them in your projections, see above Balance Sheet projections), then CFO equals net income plus the depreciation expense.

If you are getting fancy, then you will need to make further adjustments to net income to calculate the actual cash flow from operations.

Accounts Payable: Throughout the course of a month, you will pay some previously owed vendors and accrue new accounts payable. The *change* from month to month is the adjustment. As an example, if you project that at the end of January the A/P balance is $500 and the end of February the A/P balance is $700, then you had $200 *more* in expenses than you paid for in January. The increased balance of $200 suggests that, while you may have paid off some of your old debt, you still increased your unpaid balances by $200. This increase means that you have $200 more cash available in the business than the net income would suggest.

- Increasing A/P balance—Add the difference to the net income.
- Decreasing A/P balance—Subtract the difference from net income.

Accounts Receivable: Throughout the course of a month, some customers that previously owed you money will pay you, and you will have new customers that delay payment. The *change* from month to month is the adjustment. For example, if you project that at the end of January the A/R balance is $1,500 and the end of February the A/R balance is $1,200, this suggests that you received $300 *more* cash in January than your revenues would suggest. The decreased balance of $300 suggests that, while some customers delayed payments, you had more customers paying off old receivables than you had new ones. This increase means that you have $300 more cash available in the business than the net income would suggest.

- Increasing A/R balance—Subtract the difference from net income
- Decreasing A/R balance—Add the difference to the net income.

Cash Flow from Investing (CFI)

All the purchases that you made and recorded on the balance sheet are noted in the CFI section of the cash flow statement. Even if you received financing to make the purchases (which would be a cash *inflow* from financing), you still need to note the *outflow* from investing.

If you sell equipment, then the money received will also be noted in this section as an *inflow*.

Cash Flow from Financing (CFF)

Money comes in and out of your business through financing. If you receive money from a loan or investor, this is an *inflow*. If you pay down debt (such as a car payment, mortgage or operating loan), this in an *outflow*.

By noting the cash flow from financing, potential investors can see if you financed your investing purchase through your operating revenues or through a loan. If your business remains cash positive despite continued losses, the cash flow statement will show where the money is coming from. These are details that you probably know intuitively, but an investor will want to see.

This table shows Kristen's Statement of Cash Flows:

	Jan	Feb	Mar	Apr	May	Jun	Jul	Aug	Sep	Oct	Nov	Dec	Total 2013
Cash Flow From Operations													
Net Income	(726)	(264)	(4,317)	(10,486)	(13,519)	191	2,242	11,369	7,999	6,725	1,871	836	1,920
Adj for Depreciation	518	518	518	518	518	518	518	518	518	518	518	518	6,214
Total CFO	(208)	254	(3,800)	(9,968)	(13,001)	708	2,760	11,887	8,517	7,243	2,389	1,354	8,134
Cash Flow From Investing													
Mower					2,500								2,500
Box Truck					15,000								15,000
Greenhouse		8,000											8,000
Total CFI	—	(8,000)	—	—	(17,500)	—	—	—	—	—	—	—	(25,500)
Cash Flow From Financing													
Inflows													
Credit Card Loan	25,000												25,000
Grant 1		2,000											2,000
Grant 2		1,000											1,000
Loan 1				25,000									25,000
Grant 3				9,250									9,250
Total Inflow	25,000	3,000	—	34,250	—	—	—	—	—	—	—	—	62,250
Outflows													
Loan Principal Repayment	—	—	—	345	347	349	351	353	356	358	360	362	3,182
Credit Card Repayment		500	500	500	500	500	500	2,000	5,000	5,000	5,000	5,000	25,000
Total Outflow	—	500	500	845	847	849	851	2,353	5,356	5,358	5,360	5,362	28,182
Total CFF	25,000	2,500	(500)	33,405	(847)	(849)	(851)	(2,353)	(5,356)	(5,358)	(5,360)	(5,362)	34,068
Total Cash Flow	24,792	(5,246)	(4,300)	23,437	(31,348)	(141)	1,908	9,533	3,161	1,885	(2,971)	(4,009)	16,703
Beginning Cash	5,000	29,792	24,546	20,246	43,683	12,335	12,194	14,103	23,636	26,797	28,683	25,711	5,000
Ending Cash	29,792	24,546	20,246	43,683	12,335	12,194	14,103	23,636	26,797	28,683	25,711	21,703	21,703

8. Sensitivity Analysis

Throughout the process of building these projections, you've stopped to make sure everything's on track and the numbers are adding up. Presumably at this point, you're confident the numbers work and you've written a plan for a viable business.

Nonetheless, it's worth scaling back your projections one more time. Just in case. Just in case your start-up expenses are more than you expect. Just in case, it takes longer to build your customer base. Just in case any number of things that don't go as planned.

Over the years, I've worked with many clients. Something always goes wrong. Here's a list of some of things that happened.

- The grant they were hoping to get was not awarded.
- The building contractor disappeared. After a new one is found, he disappears too.
- It rains too much.

- It doesn't rain enough.
- The field has more boulders in it than expected.
- The primary growing field floods during heavy rains.
- Rodents are burrowing holes in the fields.
- The town won't let them put a sign on the highway directing potential customers to the farm store.
- The cider press took longer to restore than anticipated and a full apple season was missed.
- The zoning board won't let them build a kitchen without investing $10,000 they hadn't planned for.

For your own peace of mind, and that of your investors, test out the worst-case scenario. Go through your assumptions:
- Scale back your revenue assumptions by 15%
- Increase your start-up costs by 10%
- Increase your operating expenses by 10%

What happens to your profitability and cash flow? Do you need to borrow more money? Can you afford to borrow more money? Play around with the numbers to find the worst-case where you can still build a viable business.

9. Revisit Your Business Plan

With all the back and forth with the numbers, you may have laid out a different story of how your business will work. Go back through your business plan to make sure the story matches your numbers. If you put sales projections into your business plan, make sure they match the numbers you settled into for your projections. If you noted profit margin in the narrative, make sure you have the right number.

You may also want to copy and paste a few summary numbers:
- Summary revenue
- Summary net income
- Summary of cash flow
- Profit margin

So there you have it: financial projections in 9 easy (*ahem*) steps!

Notes

1. Net Income plus depreciation. Depreciation is a non-cash expense which means Kristen's cash from operations was $28,130. She can use the money to pay herself or reinvest in her business.
2. Need a refresher on fixed and variable costs? Refer to Chapter 2: Building the Foundation: The Financial Statements and Basic Accounting.

3. The formula entered into your calculator will be .40 × .55

4. Both employees and employers pay a portion of the payroll taxes. The employee's portion is deducted from their paycheck.

5. A word of caution about grants: there is no guarantee that you will get grants, and you should not base the success of your business on the ability to get grants.

6. irs.gov/pub/irs-pdf/p946.pdf

7. For a quick and dirty calculation, use the following formula in Excel
=PMT(rate, npr, pv); where
Rate = the interest rate divided by the number of payments you will make in a year. For example, if the annualized interest rate is 7%, and you will make 12 payments per year, the interest rate is 7%/12 or .58%
npr = The number of payments that you will make. A 5-year loan with monthly payments will be 60 periods.
PV = the amount of money you are borrowing.
Calculating the monthly payment for a 7%, 5-year loan on $10,000 would look like this: =PMT(.58%,60,10000). Note that there are no comma separators in the numbers

8. For the purposes of business planning, a business balance sheet is sufficient. You do not need to list personal assets and liabilities. However, some banks and lenders will also want to see a personal balance sheet, that is, what do you own and what do you owe that is not part of the business. This could include your personal vehicle, your house, the mortgage on your house and/or student loans.

9. With the exception of land.

10. For the purposes of these cash flow projections, we will ignore Sales Tax Payable. Depending on where you live, you may collect sales (or meals) tax for some of the goods that you sell. You receive the money from your customers and then pass it along to your state revenue department. This money is not revenue, and it does not belong to you, even though it can be in your bank account for as long as 3 months (one quarter). If you properly account for sales tax, it will not show up on your income statement, but it will on the balance sheet.

Templates

Templates used throughout the book may be accessed from:
juliashanks.com/TheFarmersOfficeTemplates/

The templates included in the zip file are (in alphabetical order):
Balance Sheet Template—Excel
Calculating Depreciation—Excel
Cash Call—Excel
COA[1] for QB[2] Upload—Livestock Farm—Excel
COA for QB Upload—Vegetable Farm—Excel
Debt Service—Excel
Executive Summary—PDF
Importing COA into QB Instructions—PDF
Income Statement Projections—Excel
Inventory Tracking Worksheet—Excel
Investor Presentation—PPT
Job Description—Word
Labor Template—Excel
Quick And Dirty Cash flow Worksheet—Livestock—Excel
Quick and Dirty Cash Flow Worksheet—Vegetable—Excel

[1] Chart of Accounts
[2] QuickBooks

Index

About the Author

JULIA SHANKS consults with food and agricultural entrepreneurs to achieve financial and operational sustainability. Working with a range of beginning and established farmers, she provides technical assistance and business coaching that empowers them to launch, stabilize and grow their ventures.

Julia launched her first business, Interactive Cuisine, in 1997. Through this innovative service, she provided informal cooking lessons in private homes and corporate team-building through cooking. She garnered local and national press—including *Bon Appétit Magazine* and *The Boston Globe*. This whet her appetite for business, and she discovered a love of excel. She shifted her career focus, combining her passions for food and numbers to help farmers and chefs build financially viable businesses.

A frequent lecturer on sustainable food, accounting and business management, Julia sits on the advisory board of Future Chefs and is the regional leader of Slow Money Boston. She received her professional chef training at the California Culinary Academy, her BA from Hampshire College and an MBA, Magna Cum Laude, from Babson College. She is the co-author of *The Farmers Market Cookbook*.

Julia lives in Cambridge, MA. To learn more about Julia and her work, visit her website: www.juliashanks.com

A Note About the Publisher

NEW SOCIETY PUBLISHERS (**www.newsociety.com**), is an activist, solutions-oriented publisher focused on publishing books for a world of change. Our books offer tips, tools, and insights from leading experts in sustainable building, homesteading, climate change, environment, conscientious commerce, renewable energy, and more — positive solutions for troubled times.

The interior pages of our bound books are printed on Forest Stewardship Council®-registered acid-free paper that is 100% post-consumer recycled (100% old growth forest-free), processed chlorine-free, and printed with vegetable-based, low-VOC inks, with covers produced using FSC®-registered stock. New Society also works to reduce its carbon footprint, and purchases carbon offsets based on an annual audit to ensure a carbon neutral footprint. For further information, or to browse our full list of books and purchase securely, visit our website at: **www.newsociety.com**

New Society Publishers
ENVIRONMENTAL BENEFITS STATEMENT

For every 5,000 books printed, New Society saves the following resources:[1]

35	Trees
3,189	Pounds of Solid Waste
3,508	Gallons of Water
4,576	Kilowatt Hours of Electricity
5,797	Pounds of Greenhouse Gases
25	Pounds of HAPs, VOCs, and AOX Combined
9	Cubic Yards of Landfill Space

[1]Environmental benefits are calculated based on research done by the Environmental Defense Fund and other members of the Paper Task Force who study the environmental impacts of the paper industry.

MIX
Paper from
responsible sources
FSC® C016245